# GRANITE

# OCCURRENCE, MINERALOGY AND ORIGIN

# EARTH SCIENCES IN THE 21ST CENTURY

Additional books in this series can be found on Nova's website under the Series tab.

Additional e-books in this series can be found on Nova's website under the e-book tab.

Earth Sciences in the 21st Century

# Granite

# Occurrence, Mineralogy and Origin

Miroslava Blasik
and
Bogdashka Hanika
Editors

Nova Science Publishers, Inc.
*New York*

Copyright © 2012 by Nova Science Publishers, Inc.

**All rights reserved.** No part of this book may be reproduced, stored in a retrieval system or transmitted in any form or by any means: electronic, electrostatic, magnetic, tape, mechanical photocopying, recording or otherwise without the written permission of the Publisher.

For permission to use material from this book please contact us:
Telephone 631-231-7269; Fax 631-231-8175
Web Site: http://www.novapublishers.com

## NOTICE TO THE READER

The Publisher has taken reasonable care in the preparation of this book, but makes no expressed or implied warranty of any kind and assumes no responsibility for any errors or omissions. No liability is assumed for incidental or consequential damages in connection with or arising out of information contained in this book. The Publisher shall not be liable for any special, consequential, or exemplary damages resulting, in whole or in part, from the readers' use of, or reliance upon, this material. Any parts of this book based on government reports are so indicated and copyright is claimed for those parts to the extent applicable to compilations of such works.

Independent verification should be sought for any data, advice or recommendations contained in this book. In addition, no responsibility is assumed by the publisher for any injury and/or damage to persons or property arising from any methods, products, instructions, ideas or otherwise contained in this publication.

This publication is designed to provide accurate and authoritative information with regard to the subject matter covered herein. It is sold with the clear understanding that the Publisher is not engaged in rendering legal or any other professional services. If legal or any other expert assistance is required, the services of a competent person should be sought. FROM A DECLARATION OF PARTICIPANTS JOINTLY ADOPTED BY A COMMITTEE OF THE AMERICAN BAR ASSOCIATION AND A COMMITTEE OF PUBLISHERS.

Additional color graphics may be available in the e-book version of this book.

**Library of Congress Cataloging-in-Publication Data**

Granite : occurrence, mineralogy, and origin / editors, Miroslava Blasik and Bogdashka Hanika.
  p. cm.
Includes bibliographical references and index.
ISBN 978-1-62081-566-3 (hardcover)
1.  Granite. 2.  Physical geology. 3.  Petrology.  I. Blasik, Miroslava. II. Hanika, Bogdashka.
QE462.G7G726 2011
553.5'2--dc23
                          2012013579

*Published by Nova Science Publishers, Inc. † New York*

# CONTENTS

| | | |
|---|---|---|
| **Preface** | | vii |
| **Chapter 1** | Mineralogy and Origin of the Rare-Earth-Element-Bearing Granites from Central India<br>*Yamuna Singh* | 1 |
| **Chapter 2** | Occurrence of Th, U, Y, Zr, and Ree-Bearing Accessory Minerals in Granites and Their Petrogenetic Significance<br>*Miloš René* | 27 |
| **Chapter 3** | Crust and Mantle Contributions to Orogenic Granitoid Magmatism: The Case of the Tertiary Magmatism of Alps (Italy)<br>*Laura Pinarelli, Angelo Peccerillo and Carmelita Donati* | 57 |
| **Chapter 4** | Contrasting Petrological Attributes of Granites in Nigeria<br>*Ademuyiwa Adetunji* | 79 |
| **Chapter 5** | Petrology, Geochemistry and Origin of Topaz Granite<br>*Miloš René* | 99 |
| **Chapter 6** | Nonlinear Acoustic Phenomena in Granite<br>*V. E. Nazarov, A.B. Kolpakov and A. V. Radostin* | 117 |
| **Chapter 7** | Experimental Methods of Determining Thermal Properties of Granite<br>*Olukayode D. Akinyemi, Thomas J. Sauer and Yemi S. Onifade* | 131 |
| **Chapter 8** | The Shoshonitic Granitoids of Altai-Sajan Folded Area: Petrology and Ore Mineralization<br>*A. I. Gusev and N. I. Gusev* | 143 |
| **Index** | | 149 |

# PREFACE

In this book, the authors gather and present topical research in the study of the occurrence, mineralogy and origin of granite. Topics discussed in this compilation include the occurrence of Th, U, Y, Zr, and REE-bearing accessory minerals in granites and their petrogenetic significance; contrasting petrological attributes of granites in Nigeria; nonlinear acoustic phenomena in granite; experimental methods of determining thermal properties of granite; and the mineralogy of the rare-earth element-bearing granites from Central India.

Chapter 1 - The rare-earth-element (REE)-bearing granitoids of the Precambrian Chhotanagpur Granite Gneiss Complex (CGGC) terrain are texturally three types (i) coarse-grained porphyritic granitoid (PRG), (ii) medium-grained (a) grey granitoid (GG) and (b) pink granitoid (PG), and (iii) fine-grained grey granitoid (FGGG). Based on the IUGS recommended parameters, the granitoids can be classified as (i) PRG: hornblende-biotite granite and biotite granite, (ii) GG: muscovite-biotite granite and muscovite granite, and (iii) FGGG and PG: biotite granite and muscovite-biotite granite, respectively. The whole-rock chemical data suggest that, similar to the modal compositions, the chemical compositions of the granitoids are also, by and large, similar. Chemically, like the classification based on modes, all the granitoids can also be classified as granite (ss). Available data indicate that the granitic melt was generated in response to deformation and extensive regional metamorphism coeval to $F_3$ deformation and $M_3$ metamorphism known from the CGGC terrain. After generation of the melt, its initial emplacement and crystallisation took place at deeper level, where an early differentiate as porphyritic granite formed. Later, a relatively more chemically evolved melt moved to a shallower level, where crystallisation of the medium-grained GG took place. The distinctly fine-grained nature and limited areal extent of the FGGG make it possibly a phase different from the PRG-GG phase. Accordingly, it may be taken to represent a localised granitic activity, unrelated to the widespread former granitic activity. As against 1005 Ma age of the GG, the distinctly younger isochron age of the pink granite (815 Ma), with signatures of metasomatism in PG and the proximity of PG to FGGG, it appears that the FGGG was, perhaps, responsible for the genesis of the PG. Accordingly, the genesis of the FGGG melt and PG might be linked with the younger metamorphic processes at ~815 Ma. Significantly, the $Na_2O/Al_2O_3$ vs $K_2O/Al_2O_3$ plots reveal that the PRG, GG, FGGG, and PG are the products of the melts derived from anatexis of crustal metasedimentary suites. The first phase of the LREE mineralisation was associated with the emplacement of the large-scale granitic bodies in the form of PRG-GG at ~1005 Ma, whereas, the second phase was linked with granitic activity at 815 Ma, represented by the FGGG and pink granite. In contrast, the heavy-rare-earth-element (HREE) mineralisation in the form of xenotime, hosted in the FGGG and PG, took place at ~815 Ma, which was accompanied by a second phase of the LREE mineralisation. In the Bastar Cratonic Terrain (BCT), limited concentrations of REE are known in granites and pegmatites and associated soils in various parts. Mineralogically, REE-bearing felsic bodies are biotite and two-mica-type monzo-granite to syeno-granite. Quartz, microcline, microcline perthite, and sodic plagioclase ($An_{5-20}$) are the major minerals, whereas, biotite and muscovite form minor minerals. Accessory minerals are apatite, zircon, sphene, allanite, topaz, fluorite, and opaques. Presence of fluorite, topaz, and cassiterite is notable, whereas, monazite and/or xenotime constitute REE-bearing phases. Geochemical data show that the REE-bearing granitic bodies are comparable in their chemical composition. They are peraluminous with normative corundum. In the Rb-Sr-Ba ternary diagram, majority of them plot in the field for strongly differentiated granite. A syn-collisional tectonic setting is indicated in Rb vs Y + Nb plots for these granites. Total REE content in the granite in general is high, with relative enrichment of LREE over HREE and pronounced negative Eu-anomaly. Early intrusive syntectonic granitic activity, involving anatexis and related migmatisation of pre-existing rocks forming gneissic granites and migmatites, took place in the Neoarchaean, for which Rb-Sr, whole-rock, isochron ages range from 2528 to 2659 Ma. Subsequently, the area witnessed the main granite tectonism in the Palaeoproterozoic, which culminated in the emplacement of rare-metal and rare-earth-bearing pegmatites. The available whole-rock, Rb-Sr ages of felsic bodies from different parts of Bastar craton suggest remobilization of sialic crust that resulted in the emplacements of REE-bearing granitic bodies from an anatectic melt during the period ca. 2300-2100 Ma.

Chapter 2 - The association of Th, U, Zr, Y and REE accessory minerals (zircon, monazite, xenotime) in two-mica granites and topaz granites of the Saxo-Danubian granite belt (Central European Hercynides) are discussed. The occurrence of these minerals is controlled by the initial trace-element contents of the melt, the aluminous saturation index, the Ca content, as well as F, P and Y contents, and the LREE/HREE and U/Th ratios of the melt. In topaz granites the late-magmatic and post-magmatic hydrothermal processes these assemblages and their compositions also control. The ca. 400 km long plutonic megastructure of the Saxo-Danubian belt is formed by Fichtelgebirge/Erzgebirge compositional batholith in the Saxothuringian Zone and the South Bohemian batholith (SBB) in the Moldanubian Zone of the Central European Hercynian fold belt. Two-mica granites form an important part of the SBB. Two, mineralogically and geochemically contrasting granites, the Eisgarn and the Deštná granites, represent the majority of these granites. In the Krušné Hory/Erzgebirge area according to their geochemical signatures P-rich and P-poor topaz granites can be distinguished. In the Eisgarn granite accessory minerals are concentrated in biotite flakes, whereas in the Deštná granite accessory minerals are predominantly enclosed in K-feldspar phenocrysts. These differences together with different Zr, Th and LREE concentrations in both granite types indicate quite different melting histories of both granites melts. Thorium, U, Y, Zr and REE-bearing accessory minerals in both two-mica granite types are represented by apatite, monazite, and zircon. In addition, in the Deštná granite rather rare xenotime-(Y) occurs. Monazite in both granite types is Ce-dominant (1.3–1.8 apfu Ce) with a restricted amount of cheralite component. However, monazite from the Deštná granite is partly enriched in Y (up to 8.6 mol.% $YPO_4$). Xenotime-(Y) from the Deštná granite has a composition along the solid solution between xenotime and cheralite with predominance of U over Th (Th/U = 0.08–0.33). For topaz granites zircon–xenotime intergrowths are significant. Zircon from the

P-rich topaz granites displays a significant enrichment in P (up to 8.3 wt.% $P_2O_5$) and depletion in Y compared to zircon from the P-poor granites. Zircon from the P-poor topaz granites contains up to 18.4 wt.% $Y_2O_3$. Xenotime from the P-poor granites also displays a considerable enrichment in HREE (up to 35.7 wt.% $HREE_2O_3$) compared to xenotime from P-rich granites (up to 19.5 wt.% $HREE_2O_3$). For xenotime from P-rich granites the brabantite-type substitution is particularly significant, whereas in xenotime from P-poor granites the thorite-type substitution occurred.

Chapter 3 - Tertiary orogenic magmatic rocks occur extensively along the Alps. Exposed lithologies are dominated by intermediate to acid granitoids but also include some dikes and volcanics. Major, trace element, and isotopic signatures show important variations along the Alps, highlighting complex origin and differentiation processes. Calcalkaline to shoshonitic magmatism is ubiquitous, whereas ultrapotassic alkaline rocks are restricted to the Western Alps. A few peraluminous granites, representing pure crustal anatectic magmas, are associated to some of the major intrusive bodies. Individual plutons, such as Bregaglia and Adamello, exhibit positive covariation of whole rock $^{87}Sr/^{86}Sr$ vs. $\delta^{18}O$, a feature indicative of magma-wall rock interaction. However, some incompatible element ratios (e.g., Th/Nb, La/Nb, Th/Y) do not change much during this process, suggesting that these ratios may be used to constrain compositions of parental magmas. Mantle normalized incompatible element patterns of mafic rocks (MgO >3 wt%) are fractionated and enriched in Large Ion Lithophile Element (LILE: Rb, K, LREE, U, Th, Pb) and relatively depleted in High Field-Strength Elements (HFSE: Ta, Nb, Zr, Hf, Ti). The lowest LILE abundances and LILE/HFSE ratios are observed in the mafic rocks from Adamello. Sr-Nd isotope ratios are negatively correlated both at the regional scale and within single granitoid bodies. The highest Sr- and lowest Nd-isotope ratios are observed in the Western Alps, where some mafic potassic alkaline dykes show crustal-like isotopic signatures. Pb-isotope ratios have more restricted ranges than Sr-Nd isotopes, and mostly fall in the compositional field of the upper crust. Low Sr- and high Nd-isotope ratios and relatively low Pb-isotope signatures are observed in some Adamello mafic rocks in the Central Alps, reflecting the contribution from relatively uncontaminated mantle sources. Overall, geochemical and isotopic evidence suggests that both crust and mantle end-members contributed to the origin and evolution of Alpine Tertiary magmatism. Some compositional features for individual plutons, such as covariation of Sr and O isotopic ratios, are related to local processes of magma contamination by crustal wall rocks. In contrast, other features such as regional distribution of petrological types and variation of LILE/HFSE and Sr-Nd-Pb isotopes of mafic rocks along the Alps, are better explained by heterogeneous mantle sources that had been variably contaminated by fluids released from subducted upper crustal material during Africa-Europe convergence. Occurrence in the Western Alps of ultrapotassic rocks with very high Sr- and low Nd-isotope ratios suggests a stronger degree of mantle metasomatism by upper crust in this sector. By contrast, lower LILE/HFSE ratios and lower Pb-isotope ratios at Adamello highlight stronger role of pristine mantle end-members, which were not contaminated by subducted crust. Such a disparity of crustal contribution to magmatism is related both to variation of local stress regimes and to modification of attitude and nature of the subducting slab during Africa-Europe collision.

Chapter 4 - In Nigeria, granites were emplaced at two different geological ages in contrasting tectonic regimes. One was emplaced during the Pan-African orogeny in the Neoproterozoic (ca. 600 Ma) while the other was associated with the Mesozoic anorogenic tectonic event which was localized to the northcentral Nigeria. The older of this is referred to as the Pan-African Granites or the Older Granites while the younger is often referred to as the Younger Granites. The Pan-African Granites represent the most prominent features of the Pan-African orogeny and are essentially pre-, syn- and post-tectonic. They occur as stocks, bosses and batholiths, and are widely distributed in the Basement Complex of Nigeria. The Pan-African Granites with other plutonic igneous rocks of the same age range are collectively called Pan-African Granitoids. The granitoids range in composition from diorites and tonalites through granodiorites to true granites and syenite. Charnockites, pegmatites and aplites are also associated with them. The Pan-African Granites constitute a major petrological unit of the Basement Complex of Nigeria. They are notable for lack of any mineralization. However, certain categories of the associated pegmatites contain economic quantity and grade of many rare metals: Sn, Nb, Ta, Li etc. and host of gemstones: tourmaline, beryl, garnet and industrial minerals like feldspars, muscovite and quartz. The Younger Granites occur as groups of ring dykes called ring complexes. They constitute parts of a larger province of mid-plate anorogenic alkaline magmatism of the Mesozoic period. They occupy a zone of about 200 km wide and extends from the Benue Valley in the south to about 320 km northwards within the country. Outside the country, the zone extends to a distance of about 1,600 km into the Republic of Niger. The ages of the granites running from the most northerly Niger range from about 470 ± 5 Ma in Air through 310 Ma in Zinda to about 141 Ma in Afu at the southernmost in Nigeria. The Younger Granites are essentially high-level intrusives and mostly peralkaline with subordinate peraluminous and metaluminous. More than 50 ring complexes have been identified. The diameters of the complexes vary from < 2 to > 25 km and individual massifs varying from about 1,000 km$^2$ to < 1 km$^2$. Four major petrographic groups of the Younger Granites are distinguished: hornblende-pyroxene-fayalite granites, hornblende-biotite-granites, biotite granites and riebeckite granites. Other associated igneous rocks are syenite, gabbros, rhyolites, dolerites, basalts, agglomerate and tuffs. The albitized riebeckite granites and albitized biotite granites are known to contain Nb, Sn, U, Li and PO$_4$ mineralization. So also, the associated minor pegmatites, pegmatitic granites, greisens and quartz veins contain Sn, Zn, Be, Pb, Cu, W and Th mineralization. The alluvial and eluvial deposits of Sn and Nb derived from these mineralized Younger Granites largely supported the economy of Nigeria up to the late 1950's when crude oil was discovered in commercial quantity.

Chapter 5 - Topaz granites are alkali feldspar granites containing albite, quartz, K-feldspar, lithium mica and topaz, which are characterized by their an extreme enrichment in F (up to 4 wt.%), enrichment in Li, Rb, Cs, and also in Sn and W. These granites are usually subdivided into P-rich and P-poor subtypes. In the Saxothuringian Zone of the Bohemian Massif (Central European Hercynides) both types occur, which are connected with economically interested Sn-W-Nb-Ta-Li mineralization. The most important occurrences of the P-rich granite suite are stocks in the area of the Ehrenfriedersdorf and Horní Slavkov–Krásno ore districts, whereas Altenberg and Cínovec/Zinnwald granite stocks represent the P-poor granite suite. The P-rich granites host large amounts of apatite, together with monazite and relatively scarce xenotime. In P-poor granites apatite is absent and the low whole-rock P content (< 0.1 wt.% P$_2$O$_5$) is reflected by variable amounts of monazite and xenotime. These granites also have a significant occurrence of thorite, Hf-highly enriched zircon with low P amount, bastnäsite, synchysite-(Y), chernovite-(Y) and complex Zr(Hf)-Th-REE phases. The P-rich granites are late-collisional, strongly peraluminous S-type granites (A/CNK = 1.1–1.5) with enrichment in P and depletion in REE and other HFSE. The distinctly high phosphorus contents of these granites (0.3–0.7 wt.% $P_2O_5$), high Al content and low Ca abundance (<0.2 wt.% CaO) facilitated the incorporation of P into the structure of late crystallized silicate minerals (albite, K-feldspar, topaz, zircon). The P-poor granites are post-collisional, slightly peraluminous granites (A/CNK = 1.1–1.2), with very low P and high REE and HFSE. These granites display some signatures of A-type granites. Contrasting patterns in the behavior of phosphorus in P-rich and P-poor granites could be explained by the strong control of peralumosity on apatite solubility in the granite melt. The significant enrichment in P of P-rich granites leads to the formation of P- and Al-bearing complexes in residual melts resulting in high P-contents of feldspars, topaz and zircon. The P-rich topaz granites were derived by anatectic melting of upper crustal, predominantly metasedimentary sequences of the Saxothuringian Zone, whereas the P-poor topaz granites were derived by small-volume partial melting of lower crustal felsic granulites and/or metagranites.

Chapter 6 - Granite as a rock with rather complex microstructure possesses a strong acoustic nonlinearity in comparing with nonlinearity of homogeneous materials such as water, glass etc. This fact is associated with the presence in their structure of different defects such as dislocations, micro-cracks, grain boundaries etc, and it was confirmed and accepted by number of researchers. This chapter contains the results of two experimental studies of nonlinear acoustic phenomena in a bar resonator made of Pitkyarant Karelian granite. The first set of experiments is the study of amplitude dependent internal friction phenomena (nonlinear loss and shift in resonance frequency) and generation of the second harmonic in the case of bar exciting in the low-frequency range at its eigenfrequencies (from 3 kHz to 17 kHz). The revealed amplitude dependences are described analytically in the framework of hysteretic state equations analogously to Granato-Lücke dislocation theory. It was shown that cubic hysteresis manifests itself at small amplitudes of strains, whereas quadratic hysteresis occurs at large ones. In the second set of experiments the weak ultrasonic pulses (with carrier frequency in the range from 150 kHz to 1 MHz) were excited simultaneously with the low-frequency wave in resonance. It was revealed that characteristics of pulses propagation (the nonlinear attenuation and carrier frequency phase delay) are dependent on low frequency wave amplitude and frequency. The observed phenomena are described analytically in terms of the phenomenological state equation containing hysteretic and dissipative nonlinearities. Also a frequency dependence of nonlinearity is revealed. The effective values of nonlinear parameters of granite are estimated. The results of the study can be used for development of nonlinear acoustic diagnostics technique for rocks.

Chapter 7 - Determination of thermal properties of granite using the block method is discussed and compared with other methods. Problems that limit the accuracy of contact method in determining thermal properties of porous media are evaluated. Thermal properties of granite are determined in the laboratory with and without application of thermal interface material TIM (Arctic Silver®) to study effects of thermal contact resistance. The thermal properties analyzer KD2 (line - source heat dissipation probe) is also used with and without TIM to measure thermal conductivity of the sample. Results from block method and KD2 analyzer, with and without TIM, are compared with standard values. Results indicate significant differences with consideration of thermal contact resistance. Thermal conductivity of the granite sample increased from 2.95 W/mK to 3.95 W/mK with the standard values for granite ranging from 2.0 and 7.0 W/mK. Volumetric heat capacity decreased from $7.17 \times 10^4$ $J/m^3K$ to $5.95 \times 10^4$ $J/m^3K$, thermal diffusivity increased from $0.412 \times 10^{-4}$ $m^2/s$ to $0.67 \times 10^{-4}$ $m^2/s$ while heat flux density increased from $2.63 \times 10^{-1}$ $W/m^2$ to $4.9 \times 10^{-1} W/m^2$.

The difference in thermal conductivities with or without TIM is significant at ($P > 0.05$) which implies the effectiveness of the thermal interface material in reducing thermal contact resistance.

Chapter 8 - In the paper presented data about shoshonitic granitoids Altai-Sajan folded area. Petrology of granitoids and link with them ore mineralization considered. Deposits of W, Mo, Ta, Nb, Au, Be, REE known in paragenesis and space link with shoshonitic granites.

*Chapter 1*

# MINERALOGY AND ORIGIN OF THE RARE-EARTH-ELEMENT-BEARING GRANITES FROM CENTRAL INDIA

## Yamuna Singh[*]

*Atomic Minerals Directorate for Exploration and Research,
Department of Atomic Energy, Government of India,
Begumpet, Hyderabad, India*

## ABSTRACT

The rare-earth-element (REE)-bearing granitoids of the Precambrian Chhotanagpur Granite Gneiss Complex (CGGC) terrain are texturally three types (i) coarse-grained porphyritic granitoid (PRG), (ii) medium-grained (a) grey granitoid (GG) and (b) pink granitoid (PG), and (iii) fine-grained grey granitoid (FGGG). Based on the IUGS recommended parameters, the granitoids can be classified as (i) PRG: hornblende-biotite granite and biotite granite, (ii) GG: muscovite-biotite granite and muscovite granite, and (iii) FGGG and PG: biotite granite and muscovite-biotite granite, respectively. The whole-rock chemical data suggest that, similar to the modal compositions, the chemical compositions of the granitoids are also, by and large, similar. Chemically, like the classification based on modes, all the granitoids can also be classified as granite (ss).

Available data indicate that the granitic melt was generated in response to deformation and extensive regional metamorphism coeval to $F_3$ deformation and $M_3$ metamorphism known from the CGGC terrain. After generation of the melt, its initial emplacement and crystallisation took place at deeper level, where an early differentiate as porphyritic granite formed. Later, a relatively more chemically evolved melt moved to a shallower level, where crystallisation of the medium-grained GG took place. The distinctly fine-grained nature and limited areal extent of the FGGG make it possibly a phase different from the PRG-GG phase. Accordingly, it may be taken to represent a localised granitic activity, unrelated to the widespread former granitic activity. As against 1005 Ma age of the GG, the distinctly younger isochron age of the pink granite (815 Ma), with signatures of metasomatism in PG and the proximity of PG to FGGG, it appears that

---

[*] E-mail: yamunasingh2002@yahoo.co.uk.

the FGGG was, perhaps, responsible for the genesis of the PG. Accordingly, the genesis of the FGGG melt and PG might be linked with the younger metamorphic processes at ~815 Ma. Significantly, the $Na_2O/Al_2O_3$ vs. $K_2O/Al_2O_3$ plots reveal that the PRG, GG, FGGG, and PG are the products of the melts derived from anatexis of crustal metasedimentary suites. The first phase of the LREE mineralisation was associated with the emplacement of the large-scale granitic bodies in the form of PRG-GG at ~1005 Ma, whereas, the second phase was linked with granitic activity at 815 Ma, represented by the FGGG and pink granite. In contrast, the heavy-rare-earth-element (HREE) mineralisation in the form of xenotime, hosted in the FGGG and PG, took place at ~815 Ma, which was accompanied by a second phase of the LREE mineralisation.

In the Bastar Cratonic Terrain (BCT), limited concentrations of REE are known in granites and pegmatites and associated soils in various parts. Mineralogically, REE-bearing felsic bodies are biotite and two-mica-type monzo-granite to syeno-granite. Quartz, microcline, microcline perthite, and sodic plagioclase ($An_{5-20}$) are the major minerals, whereas, biotite and muscovite form minor minerals. Accessory minerals are apatite, zircon, sphene, allanite, topaz, fluorite, and opaques. Presence of fluorite, topaz, and cassiterite is notable, whereas, monazite and/or xenotime constitute REE-bearing phases.

Geochemical data show that the REE-bearing granitic bodies are comparable in their chemical composition. They are peraluminous with normative corundum. In the Rb-Sr-Ba ternary diagram, majority of them plot in the field for strongly differentiated granite. A syn-collisional tectonic setting is indicated in Rb vs Y + Nb plots for these granites. Total REE content in the granite in general is high, with relative enrichment of LREE over HREE and pronounced negative Eu-anomaly.

Early intrusive syntectonic granitic activity, involving anatexis and related migmatisation of pre-existing rocks forming gneissic granites and migmatites, took place in the Neoarchaean, for which Rb-Sr, whole-rock, isochron ages range from 2528 to 2659 Ma. Subsequently, the area witnessed the main granite tectonism in the Palaeoproterozoic, which culminated in the emplacement of rare-metal and rare-earth-bearing pegmatites. The available whole-rock, Rb-Sr ages of felsic bodies from different parts of Bastar craton suggest remobilization of sialic crust that resulted in the emplacements of REE-bearing granitic bodies from an anatectic melt during the period ca. 2300-2100 Ma.

## INTRODUCTION

Rapidly growing demand for rare-earth-elements (REEs) is because of enormous applications of these metals in diversified fields (Hedrick, 1985; Preinfalk and Morteani, 1989; Gupta, 1993; Gupta and Krishnamurthy, 1990). Diversified applications of REEs are metallurgical (37%), petroleum catalyst (31%), ceramics and glass (29%), nuclear and other fields including superconductivity (3%). Sharp increase in demand for yttrium and heavy REEs (HREEs) particularly in the fields of advanced ceramics and superconductors have made these elements "commodities of the future" and their minerals "the minerals of high tech age" (Lottermoser, 1991; Spooner et al. 1991).

Systematic surveys and exploration for atomic minerals by the Atomic Minerals Directorate (AMD) for Exploration and Research, intermittently, over nearly last five decades have resulted in locating economically viable and commercially exploitable REE-bearing mineral resources in various parts of central India. Significantly, among them, bulk of the rare-earth-element and yttrium (REE and Y) mineral deposits are generally associated with granitic rocks and related soils. In this article, mineralogy and origin of rare-earth-bearing granites of central India have been briefly described. The rare-earth-bearing minerals are hosted mainly in felsic bodies in various parts of central India. Such felsic bodies are mostly located in Chhotanagpur Granite Gneiss Complex (CGGC) and Bastar cratonic terrains. Accordingly, geological setting, mineralogy, and origin of REE-bearing granitoids of both the terrains are given separately.

# GEOLOGICAL SETTING

## Chhotanagpur Granite Gneiss Complex (CGGC) Terrain

The Precambrian Chhotanagpur granite gneiss complex (CGGC) terrain covers >80,000 $km^2$-area in parts of West Bengal, Bihar, Jharkhand, Uttar Pradesh, Madhya Pradesh, and Orissa. It is dominated by granitoid gneisses and migmatites, with enclaves of different sizes, composition, and metamorphic grades occurring within the gneisses. A major part of the terrain attained upper amphibolite facies metamorphism and the rest greenschist facies. The gneissic rocks are interpreted to have formed by ultrametamorphism, whereas, the granitoids are believed to have formed from regional metamorphism and palingenetic melts (Dey, 1983; Mazumdar, 1988; Singh, 1992; Mahadevan, 2002; Singh and Reddy, 2008; Singh and Krishna, 2009).

In the western parts of the CGGC terrain, the schists and gneisses are migmatised, and have undergone multiphase tectonic deformation. In the Belangi-Jhapar area, these are intruded by distinct phases of granites, pegmatites, and vein quartz. The pegmatites occur in swarms, and are lenticular, pod-shaped, and flat bodies, with 5-300 m in length and <1 to 50 m in breadth (Singh, 1991). The presence of colourless to milky-white quartz, white to flesh-coloured blocky perthite, common graphic intergrowth of quartz and microcline, coarse columnar greenish beryl, blocky tantalite, brown mica, jet black and striated tourmaline and lack of albitisation point out that the pegmatites are of 'unevolved nature' belonging to early granitic phase (cf. Cerny, 1989; Kuster, 1990). Such features are characteristic of beryl-type pegmatites and are known to crystallize in the temperature range of 500-650$^0$C and 2 kb pressure (Cerny, 1992). The Dhabi-Bhandaria region, which also forms western part of CGGC terrain, comprises several types of granites and related pegmatites and quartz vein, with large enclaves of older metasedimentary rocks. The granites are mainly grey and pink types (Saxena et al. 1992), and are the sources of rare-earth-bearing placers in parts of the Mahan and Kanhar river basins (Singh, 1990; Rai and Banerjee, 1995; Singh and Singh, 1996; Ramesh Babu et al. 1998).

In and around Raikera-Kunkuri area, the oldest litho-units are represented by the metasedimentary sequence consisting of quartzites, phyllites, and chlorite-, mica-, and hornblende-schists. Metamorphosed basic flows (epidiorites, amphibolites, and talc-chlorite schist) are interbedded with the metasedimentary sequence. Texturally-different types of granitoids, e. g., (i) coarse-grained porphyritic granitoid, (ii) medium-grained (a) grey granitoid and (b) pink granitoid, and (iii) fine-grained grey granitoid, besides pegmatoid and quartz veins, intrude the older litho-units. At places like Kunkuri and Raikera, a gradual transition from granites to pegmatitic granites and normal and zoned pegmatites is distinctly seen. The pegmatoid veins, genetically related to the FGGG and the PG, fill fractures and joints within the parent granitoids, and are <1 to 2 m-long and <0.50 m-wide. In contrast, the pegmatites related to the GG are upto 200 m-long and 50-60 m-wide. Several pegmatites of homogeneous and zoned types, genetically related to GG, occur in the periphery of the Burha Pahar massif of the GG, especially on the southern and eastern margins. These pegmatites have been mined for beryl. Pegmatites which are present within the grey granitoids either show transitional nature, or fill the fractures and joints.

## Bastar Cratonic Terrain (BCT)

The Bastar Cratonic Terrain (BCT) exposes a well-developed sequence of rocks ranging in age from Archaean to Neoproterozoic. It forms the central part of the Indian Peninsular Shield occupying an area of 40,000 km$^2$, and is flanked by the Chhattisgarh basin in the north, Eastern Ghats khondalite-charnockite complex in the south and east and the Pranhita-Godavari basin in the southwest. In BCT, rare-metal and rare-earth (RMRE) pegmatites are located over 80 km in length and 15-20 km in width, in a general NW-SE direction.

The rocks present in the BCT are low- to medium-grade metasedimentary rocks and metabasic intrusive, all belonging to the Bengpal Group of Archaean age. The metasedimentary rocks are represented by quartzite, quartz-sericite schist, andalusite-sericite schist, staurolite gneiss, and biotite-cordierite gneiss (Crookshank, 1963; Mishra et al. 1988). The quartzitic rocks usually trend parallel to the foliation planes of schistose rocks along NW-SE to NNW-SSE direction and dip with low angles towards NE and ENE. The metasedimentary rocks contain quartz, feldspar, muscovite, garnet, chlorite, andalusite, staurolite, apatite, zircon, magnetite, ilmenite, and rutile in different proportions. Fine- to medium- to coarse-grained, melanocratic, massive and occasionally foliated amphibolites, metadolerites (metagabbro), and seldom metadiorites represent the metabasic rocks, which are usually emplaced as sill-like bodies and seldom cut across the schistosity planes of the metasedimentary rocks. The exposures range in size from 10 to 300 m in width and 100 m to 5 km in length and form linear ridges. The predominant trend is in a NW-SE direction with north-easterly dips. Subaqueous volcanic effusive with pillow structures have been described from Katekalyan area by Babu (1985 and 1994).

## MINERALOGY AND PETROCHEMISTRY OF REE-BEARING GRANITES

## Chhotanagpur Granite Gneiss Complex (CGGC) Terrain

*Porpyritic Granite (PRG):* The porphyritic granitoid (PRG) occurs as a sheet-like body in low-lying and plain areas around Ginabahar-Rajauti, and shows negligible variation in its mode of occurrence and texture. Megascopically, the PRG is a mesocratic and coarse-grained porphyritic rock. It exhibits fairly distinct foliation defined by alternate layers of large K-feldspar megacrysts and medium- to coarse-grained quartzofeldspathic and ferromagnesian minerals. Megacrysts of K-feldspar are often lensoidal, and the ferromagnesian minerals wrap around them to give rise to augen structure. Under the microscope, the PRG is medium- to coarse-grained, and shows grain size variation from two to seven mm, the average grain size being about three millimeters. Quartz, potash feldspar, and plagioclase constitute the essential minerals (Table 1). Quartz is normally interstitial and anhedral and, at places, occurs as rounded inclusions in potash feldspar, and elsewhere, as intergrowths with plagioclase. Undulatory extinction due to strain effects are noticed in some grains. Potash feldspar is represented by microcline and intergrowth textures by myrmekites, sieve quartz, and perthites of patchy and vein types. Plagioclase is albite-oligoclase in composition, shows polysynthetic twinning, and displays development of myrmekite intergrowths along margins. Hornblende and biotite are the varietal accessory minerals observed. Hornblende is pleochroic in shades of green. Biotite shows strong absorption from deep brown to yellowish brown, perfect cleavage, and straight extinction. It is invariably associated with hornblende with transitional nature, and inclusions of accessory minerals occur in both hornblende and biotite. Magnetite, sphene, apatite, zircon, allanite, and monazite are the ore minerals segregated within hornblende and biotite. The ore minerals are fine-grained, euhedral to subhedral, and occur as disseminated grains. Both hornblende and biotite display pleochroic haloes due to zircon inclusions within them. Allanite is euhedral and displays radiation cracks in the minerals surrounding it. Based on the modal Q-A-P diagram, content of accessory minerals, and texture, the porphyritic granite could be classified as 'hornblende-biotite granite' and 'biotite granite'.

Chemically, the PRG has $SiO_2$ varying from 63.31 to 72.42% and $Al_2O_3$ from 14.0 to 16.0%, with A/CNK ratios from 1.0 to 1.24 (Singh, 1992). It is fairly rich in $K_2O$ (av. 4.35%) and $Na_2O$ (av. 3.05%), with $K_2O/Na_2O$ ratios in the range of 1.12 to 1.70. Total iron as FeO ranges from 2.0 to 5.35%, with low MgO (av. 0.8%) and high CaO (av. 2.09%) contents. Average $Fe_2O_3/FeO$ (0.76) and $Fe_2O_3/MgO$ (2.16) ratios are the lowest, when compared with other types of granitoids. The $K_2O$ vs. $Na_2O$ plots (fields after Harpum, 1963) of the porphyritic granite reveal its transitional nature from adamellite to granite. The $Na_2O/Al_2O_3$ vs. $K_2O/Al_2O_3$ plots of the PRG lie in the sedimentary/metasedimentary field of Garrels and Mackenzie (1971).

**Table 1. Modal (Volume %) composition of the porphyritic granite (PRG) from the CGGC terrain, central India**

| Sample | Quartz | K-feldspar | Plagioclase | Biotite | Hornblende | Others[*] |
|---|---|---|---|---|---|---|
| 1 | 29.5 | 29.7 | 28.7 | 13.1 | 5.3 | 3.7 |
| 2 | 37.2 | 31.2 | 19.1 | 11.3 | - | 1.1 |
| 3 | 27.0 | 36.0 | 21.1 | 13.7 | 0.5 | 1.7 |
| 4 | 35.4 | 28.4 | 19.3 | 13.3 | 1.5 | 2.1 |
| 5 | 32.9 | 26.5 | 22.6 | 10.5 | 5.0 | 2.5 |
| 6 | 45.0 | 24.2 | 23.5 | 6.0 | - | 1.3 |
| 7 | 30.1 | 29.9 | 27.2 | 11.0 | - | 1.8 |
| 8 | 16.7 | 43.4 | 27.7 | 8.0 | 1.2 | 3.0 |
| 9 | 20.6 | 43.3 | 27.3 | 7.0 | - | 1.8 |
| 10 | 20.3 | 35.0 | 23.8 | 18.9 | - | 2.0 |
| 11 | 16.3 | 23.7 | 38.0 | 20.3 | - | 1.7 |
| Average | 28.3 | 31.9 | 25.3 | 12.1 | - | 2.1 |

[*] Includes magnetite, sphene, monazite, apatite, and zircon.

**Table 2. Modal (Volume %) composition of the medium-grained grey granite (GG) from the CGGC terrain, central India**

| Sample | Quartz | K-feldspar | Plagioclase | Biotite | Muscovite | Others[*] |
|---|---|---|---|---|---|---|
| 1 | 28.6 | 34.9 | 24.6 | 4.8 | 5.7 | 1.4 |
| 2 | 31.3 | 32.4 | 25.7 | 5.7 | 3.1 | 1.8 |
| 3 | 28.2 | 32.1 | 25.4 | 5.6 | 7.5 | 1.2 |
| 4 | 27.1 | 33.5 | 26.4 | 6.6 | 3.8 | 2.6 |
| 5 | 27.4 | 38.9 | 22.5 | 6.2 | 3.0 | 2.0 |
| 6 | 27.2 | 36.0 | 23.5 | 7.4 | 3.8 | 2.1 |
| 7 | 29.9 | 31.7 | 24.8 | 7.5 | 3.7 | 2.4 |
| 8 | 27.3 | 33.9 | 24.4 | 7.7 | 4.5 | 2.2 |
| 9 | 29.6 | 33.0 | 24.0 | 5.0 | 6.9 | 1.5 |
| 10 | 31.0 | 32.6 | 20.8 | 9.8 | 4.8 | 1.0 |
| 11 | 22.4 | 35.6 | 31.7 | 8.4 | 1.1 | 0.8 |
| 12 | 27.5 | 31.0 | 30.1 | 8.0 | 2.9 | 0.5 |
| 13 | 27.2 | 39.7 | 19.1 | 5.4 | 7.8 | 0.8 |
| 14 | 29.0 | 35.6 | 25.6 | 4.4 | 4.8 | 0.6 |
| 15 | 32.6 | 34.0 | 20.3 | 6.8 | 5.7 | 0.6 |
| 16 | 26.3 | 37.4 | 25.6 | 7.0 | 2.6 | 0.8 |
| 17 | 30.2 | 40.8 | 19.2 | 7.5 | 1.5 | 0.8 |
| 18 | 13.6 | 36.8 | 12.0 | 22.2 | 4.3 | 0.9 |
| 19 | 30.7 | 32.8 | 23.4 | 5.9 | 6.5 | 0.4 |
| 20 | 39.8 | 30.2 | 18.7 | 8.0 | 2.3 | 1.0 |
| 21 | 33.4 | 27.9 | 25.7 | Traces | 12.8 | 0.2 |
| 22 | 29.3 | 38.7 | 15.2 | 6.5 | 9.5 | 0.8 |
| 23 | 30.6 | 31.7 | 18.3 | 7.8 | 10.2 | 1.4 |
| 24 | 25.7 | 33.2 | 26.7 | 8.4 | 5.0 | 1.0 |
| 25 | 24.5 | 37.6 | 24.8 | 6.7 | 5.6 | 0.8 |
| Average | 28.4 | 34.6 | 22.4 | 8.0 | 5.5 | 0.8 |

[*] Includes monazite, apatite, and zircon.

*Grey Granite (GG):* The grey granitoid (GG) is widespread in the investigated area, and constitutes large-scale batholithic bodies and hill ranges. It is a leucocratic to mesocratic rock, and tends to be of the two-mica granite type. Under the microscope, it is medium-grained and shows hypidiomorphic granular texture. The grain size varies from two to four mm, with average grain-size being about 2.5 millimeters.

In the GG also, the essential mineral constituents are quartz, potash feldspar, and plagioclase (Table 2), whereas, muscovite and biotite occur as accessories. The development of patches and veins of perthitic intergrowths is also observed in microcline, besides twinned microcline. Quartz is normally anhedral and interstitial, and also occurs as sieve quartz in feldspars, and as intergrowths with plagioclase. The plagioclase is albite-oligoclase in composition, euhedral with corroded margins, and is altered to sericite, at places. Myrmekites are also commonly associated with it. Biotite is partly bleached, chloritised, and contains numerous pleochroic haloes in it. Muscovite is the varietal accessory mineral, constituting 1 to 10 volume %, with coarse grains, euhedral, and flaky nature, and has corroded margins with quartz. Zircon, apatite, and monazite are the other accessory minerals present. Apatite (about 10% in one sample and 0.3% in another) occurs segregated with biotite. Zircon occurs as inclusions in biotite, which is responsible for the development of pleochroic haloes in the latter. Hornblende is conspicuous by its absence in the GG. Based on the Q-A-P plots, average accessory minerals content, and texture, the GG may be named, in general, as 'muscovite-biotite granite'. Occasionally, the GG with leucocratic appearance, without change in essential mineral assemblage and texture, with sporadic shows of alteration of plagioclase, contains muscovite upto 12.8 volume%. Sericite, zircon, and biotite occur in trace quantity as the other accessory minerals in it. Accordingly, it may be specifically named as 'muscovite granite'. The absence of hornblende, sphene, and magnetite, the reduced abundance of biotite (av. 8.0 volume%), and the presence of primary muscovite, are the noticeable variations observed in the GG, relative to the PRG.

The chemical compositions of the grey granite (Table 3) reveal that it is silicic (71.59-72.88, av. 72.17% $SiO_2$) and alumina-rich (13.97-14.96, av. 14.64% $Al_2O_3$), with A/CNK ratios of 1.02-1.35 (av. 1.24). The $K_2O$ (5.10-5.25, av. 5.15%) of the GG is higher, and $Na_2O$ (2.36-2.69, av. 2.55%) is lower, than in the PRG. Likewise, the $K_2O/Na_2O$ ratios (1.90-2.22, av. 2.03) of the GG are also higher than in the PRG. Total iron (av. $Fe_2O_3$ 0.94 and FeO 0.98), MgO (av. 0.29%), and CaO (1.3%) are low, whereas, $Fe_2O_3/FeO$ (av. 0.96) and $Fe_2O_3/MgO$ (av. 3.31) ratios are high, when compared with similar ratios of the PRG. All the plots of $K_2O$ vs. $Na_2O$ of the grey granite fall in only one field, i.e., the granite. Significantly, the $Na_2O/Al_2O_3$ vs. $K_2O/Al_2O_3$ plots of the GG also lie in the same field in which all the respective plots of the PRG lie, i. e., sedimentary/metasedimentary field of Garrels and Mackenzie (1971). The grey granite has yielded an isochron corresponding to an age of 1005±51 Ma, with an initial $^{87}Sr/^{86}Sr$ ratio of 0.7047 (Singh and Krishna, 2009).

**Table 3. Chemical composition (Wt. %) of the representative medium-grained grey granites (GG) from the CGGC terrain, central India**

| Oxide/Ratio | 1 | 2 | 3 | 4 | 5 | 6 | 7 | 8 | Av. |
|---|---|---|---|---|---|---|---|---|---|
| $SiO_2$ | 71.59 | 72.24 | 72.87 | 71.90 | 72.06 | 72.88 | 72.06 | 71.75 | 72.17 |
| $Al_2O_3$ | 14.96 | 14.96 | 14.96 | 14.56 | 14.56 | 13.97 | 14.56 | 14.56 | 14.64 |
| $Fe_2O_3$ | 0.95 | 0.92 | 0.89 | 0.91 | 0.85 | 1.32 | 0.89 | 0.76 | 0.94 |
| FeO | 0.85 | 1.01 | 0.94 | 0.79 | 1.04 | 1.26 | 0.86 | 1.12 | 0.98 |
| MgO | 0.26 | 0.26 | 0.26 | 0.34 | 0.34 | 0.26 | 0.26 | 0.34 | 0.29 |
| CaO | 1.58 | 1.09 | 1.09 | 0.97 | 1.09 | 1.21 | 1.09 | 2.55 | 1.3 |
| $Na_2O$ | 2.36 | 2.36 | 2.53 | 2.53 | 2.53 | 2.69 | 2.69 | 2.69 | 2.55 |
| $K_2O$ | 5.25 | 5.10 | 5.10 | 5.25 | 5.25 | 5.10 | 5.10 | 5.10 | 5.15 |
| $TiO_2$ | 0.18 | 0.27 | 0.29 | 0.21 | 0.23 | 0.27 | 0.16 | 0.18 | 0.22 |
| $P_2O_5$ | 0.36 | 0.43 | 0.30 | 0.30 | 0.30 | 0.19 | 0.19 | 0.26 | 0.29 |
| MnO | 0.004 | 0.003 | 0.004 | 0.004 | 0.004 | 0.006 | 0.004 | 0.004 | 0.004 |
| $H_2O$ | 0.08 | 0.12 | 0.08 | 0.06 | 0.03 | 0.04 | 0.09 | 0.03 | 0.07 |
| *Total* | *99.42* | *98.76* | *98.31* | *97.82* | *98.28* | *99.20* | *97.95* | *99.34* | *98.60* |
| $K_2O/Na_2O$ | 2.22 | 2.16 | 2.02 | 2.08 | 2.08 | 1.90 | 1.90 | 1.90 | 2.03 |
| $Fe_2O_3/FeO$ | 1.12 | 0.91 | 0.95 | 1.15 | 0.82 | 1.05 | 1.03 | 0.68 | 0.96 |
| $Fe_2O_3/MgO$ | 3.65 | 3.54 | 3.42 | 2.68 | 2.50 | 5.08 | 3.42 | 2.24 | 3.31 |
| A/CNK | 1.23 | 1.35 | 1.30 | 1.29 | 1.27 | 1.17 | 1.26 | 1.02 | 1.24 |

**Table 4. Modal (Volume %) composition of the fine-grained grey granite (FGGG) from the CGGC terrain, central India**

| Sample | Quartz | K-feldspar | Plagioclase | Biotite | Others* |
|---|---|---|---|---|---|
| 1 | 29.7 | 37.1 | 23.7 | 7.7 | 1.8 |
| 2 | 36.9 | 32.8 | 22.4 | 7.6 | 0.3 |
| 3 | 37.8 | 35.7 | 22.4 | 3.7 | 0.4 |
| 4 | 32.3 | 36.8 | 24.8 | 5.0 | 1.1 |
| 5 | 33.5 | 41.0 | 18.6 | 6.5 | 0.4 |
| 6 | 27.5 | 38.4 | 28.1 | 4.2 | 1.8 |
| 7 | 36.2 | 31.6 | 21.8 | 8.6 | 1.8 |
| 8 | 39.3 | 39.6 | 18.4 | 1.9 | 0.8 |
| 9 | 33.6 | 42.9 | 22.0 | 1.0 | 0.5 |
| 10 | 31.8 | 29.9 | 27.5 | 9.2 | 1.6 |
| 11 | 29.8 | 41.7 | 22.0 | 4.2 | 2.3 |
| 12 | 27.9 | 37.3 | 23.1 | 10.7 | 1.0 |
| 13 | 29.7 | 38.9 | 28.3 | 2.3 | 0.8 |
| Average | 32.8 | 37.2 | 23.3 | 5.6 | 1.1 |

*Includes zircon, monazite, and xenotime.

*Fine-Grained Grey Granite (FGGG):* The outcrops of the FGGG are limited. It is leucocratic to mesocratic, distinctly fine-grained in the field, as compared to the other variants. Under the microscope, the FGGG is fine- to medium-grained, and shows hypidiomorphic granular texture. The grain size varies from 0.5 to 1.5 mm, with average grain size being about one millimeter. Perthites of patchy type and string type are common. Myrmekite and sieve quartz are the other intergrowth textures observed. Like the other variants, the FGGG also has quartz, microcline, perthite, and plagioclase as the essential mineral assemblage (Table 4), which does not display any noticeable difference in optical characters, mineral intergrowths, and mode of occurrence, including nature and type of inclusions. Biotite ranges from 1.0 to 10.7 volume% and shows strong absorption from deep brown to yellowish brown, perfect basal cleavage, and straight extinction. Zircon, monazite, and xenotime in traces are the accessory minerals. They are fine-grained, euhedral to subhedral, and occur scattered in the intergranular spaces and also in intimate association with biotite. Pleochroic haloes are present around inclusions of zircon in biotite. Mineralogically and texturally, the FGGG can be classified mostly as 'biotite granite'. The fine-grained nature, limited areal extension, lesser biotite content (av. 5.6 volume %), and the presence of xenotime are the distinguishing features of the FGGG relative to the other granitoids.

Whole-rock analysis of the fine-grained grey granite (Singh and Singh, 1996) reveals that, compared to other variants, it has the highest silica (av. 73.35% $SiO_2$), but like other variants, it is also alumina-rich (13.92-17.52, av. 14.79% $Al_2O_3$), with A/CNK ratios of 1.05-1.63 (av. 1.25). The $K_2O$ (4.80-6.00, av. 5.05%) and $Na_2O$ (2.40-3.40, av. 2.98%) contents are noteworthy, including $K_2O/Na_2O$ ratios (1.17-2.0, av. 1.71). Iron (av. $Fe_2O_3$ 0.72 and FeO 0.85), MgO (av. 0.35%), and CaO (0.91%) contents are low, whereas, $Fe_2O_3$/FeO (av. 0.85) and $Fe_2O_3$/MgO (av. 2.59) ratios are lower, when compared with similar ratios of the PRG and GG. As in the case of GG, all the plots of $K_2O$ vs. $Na_2O$ of the FGGG also fall in only one field, i.e., the granite. Likewise, the $Na_2O/Al_2O_3$ vs. $K_2O/Al_2O_3$ plots of the FGGG also lie in the sedimentary/metasedimentary field, similar to the plots of the PRG and GG.

*Pink Granite (PG):* The exposures of the pink granite are also limited. From the other granitoids, it distinguishes itself by its pink to buff colours. Granitic soils in the vicinity of FGGG and PG outcrops contain xenotime. Examination of thin sections under microscope reveals it to be a medium-grained rock with hypidiomorphic texture.

**Table 5. Modal (Volume %) composition of the medium-grained pink granite (PG) from the CGGC terrain, central India**

| Sample | Quartz | K-feldspar | Plagioclase | Biotite | Muscovite | Others[*] |
|---|---|---|---|---|---|---|
| 1 | 29.3 | 27.6 | 30.3 | 5.5 | 5.8 | 1.5 |
| 2 | 30.2 | 31.9 | 18.6 | 5.1 | 11.5 | 2.7 |
| 3 | 26.5 | 36.6 | 24.8 | 7.2 | 3.4 | 1.5 |
| 4 | 28.3 | 32.4 | 24.8 | 6.9 | 5.0 | 2.6 |
| 5 | 24.3 | 37.2 | 30.4 | 3.3 | 3.7 | 1.1 |
| 6 | 27.9 | 38.7 | 25.7 | 4.6 | 1.9 | 1.2 |
| 7 | 20.1 | 33.2 | 32.2 | 10.1 | 1.2 | 3.2 |
| 8 | 27.5 | 33.3 | 25.9 | 4.9 | 6.6 | 1.8 |
| 9 | 38.3 | 30.0 | 20.0 | 5.0 | 6.0 | 0.7 |
| Average | 28.0 | 33.4 | 25.9 | 5.8 | 5.0 | 1.8 |

[*] Includes zircon, monazite, and xenotime.

**Table 6. Chemical composition (wt. %) of the medium-grained pink granite (PG) from the CGGC terrain, central India**

| Oxide/Ratio | 1 | 2 | 3 | 4 | 5 | 6 | 7 | 8 | 9 | Av. |
|---|---|---|---|---|---|---|---|---|---|---|
| $SiO_2$ | 72.58 | 72.72 | 71.50 | 72.06 | 71.89 | 72.37 | 71.06 | 72.06 | 72.56 | 72.08 |
| $Al_2O_3$ | 16.24 | 14.96 | 13.83 | 14.17 | 13.83 | 14.56 | 13.63 | 14.56 | 13.63 | 14.37 |
| $Fe_2O_3$ | 1.03 | 0.91 | 1.22 | 0.96 | 0.90 | 0.78 | 1.25 | 0.71 | 0.58 | 0.92 |
| FeO | 1.16 | 0.76 | 1.47 | 1.04 | 1.37 | 0.65 | 1.36 | 0.97 | 1.04 | 1.09 |
| MgO | 0.37 | 0.26 | 0.46 | 0.34 | 0.38 | 0.26 | 0.45 | 0.34 | 0.48 | 0.37 |
| CaO | 0.08 | 1.09 | 1.09 | 1.09 | 0.83 | 1.09 | 1.40 | 1.09 | 1.01 | 0.97 |
| $Na_2O$ | 2.80 | 2.36 | 2.67 | 2.53 | 2.67 | 2.69 | 2.90 | 2.69 | 2.53 | 2.65 |
| $K_2O$ | 5.00 | 5.56 | 5.64 | 5.25 | 5.64 | 5.25 | 6.00 | 5.25 | 5.93 | 5.50 |
| $TiO_2$ | 0.24 | 0.20 | 0.26 | 0.21 | 0.25 | 0.16 | 0.31 | 0.18 | 0.27 | 0.23 |
| $P_2O_5$ | 0.37 | 0.38 | 0.15 | 0.19 | 0.12 | 0.30 | 0.18 | 0.30 | 0.07 | 0.23 |
| MnO | 0.05 | 0.003 | 0.05 | 0.003 | 0.04 | 0.003 | 0.05 | 0.004 | 0.02 | 0.024 |
| $H_2O$ | 0.06 | 0.13 | 0.02 | 0.06 | 0.05 | 0.09 | 0.14 | 0.04 | 0.20 | 0.08 |
| *Total* | *99.98* | *99.33* | *98.36* | *97.90* | *97.97* | *98.20* | *98.73* | *98.19* | *98.32* | *98.45* |
| $K_2O/Na_2O$ | 1.79 | 2.36 | 2.11 | 2.08 | 2.11 | 1.95 | 2.07 | 1.95 | 2.34 | 2.08 |
| $Fe_2O_3/FeO$ | 0.89 | 1.20 | 0.83 | 0.92 | 0.66 | 1.20 | 0.92 | 0.73 | 0.56 | 0.88 |
| $Fe_2O_3/MgO$ | 2.78 | 3.50 | 2.65 | 2.82 | 2.37 | 3.00 | 2.78 | 2.09 | 1.20 | 2.58 |
| A/CNK | 1.60 | 1.29 | 1.11 | 1.23 | 1.15 | 1.24 | 0.99 | 1.24 | 1.10 | 1.22 |

The constituent minerals are generally two to four mm in size. Although the bulk mineralogy (Table 5) and textural intergrowth of minerals of the PG, including accessory biotite with strong absorption, are not much different from what has been described earlier for the GG and FGGG, it has certain distinguishing features. Indeed, feldspars show apparent alterations to sericite, with the alkali feldspars relatively less altered, in comparison to plagioclase. In addition, biotite is chloritised. Locally, the pink granites are dominated by alkali feldspars, some grains of which are fresh, whereas, others are moderately sericitised and kaolinised. Plagioclase is commonly clouded, due to its alteration to sericite and/or muscovite. Such alterations are the manifestations of metasomatic and attendant hydrothermal activity. Quartz normally occurs as rounded inclusions in feldspars. Zircon, xenotime, and monazite constitute the heavy mineral assemblage, which occur as segregations within and along the margins of biotite. In fact, xenotime is bi-pyramidal and occurs as inclusion within the biotite. Interestingly, xenotime grains recovered from the riverine placers occurring in the vicinity are also bi-pyramidal (Rajasekharan et al., 1994), confirming interpretation that the xenotime placers of the Raikera-Kunkuri area have been derived from the specific granitoids present in the vicinity (Singh, 1990; Singh and Singh, 1996). Based on the IUGS recommended parameters (Streckeisen, 1976), average accessory modal mineralogy, and texture, the pink granite, in general, is classified as 'muscovite-biotite granite'. To sum up, the pink colour, apparent imprints of metasomatic/hydrothermal activities as reflected by altered feldspars, and relatively more abundance of xenotime (Singh and Singh, 1996), set the PG apart from the other granitoids of the area. Significantly, the presence of xenotime only in FGGG and PG could be due to some genetic relationship between the two types.

Chemical data of the pink granite (Table 6) reveal compositional similarity of certain constituents with the GG and others with the FGGG. The silica (av. 72.08% $SiO_2$), $Fe_2O_3$ (av. 0.92%), and FeO (av. 1.09%) contents are very close to those of the GG. In contrast, MgO (av. 0.37%) and CaO (av. 0.97%) and $Fe_2O_3$/MgO ratio (av. 2.58) are similar to those of the FGGG. Nevertheless, the PG has the highest $K_2O$ (av. 5.50%) content compared to the other types. The PG is also alumina-rich (av. 14.37% $Al_2O_3$), with A/CNK ratios of 0.99-1.60 (av. 1.22). The $K_2O/Na_2O$ ratios (1.79-2.36, av. 2.08) of the PG are the highest among all the investigated granites. Like other investigated granites, all the plots of $K_2O$ vs. $Na_2O$ of the PG also fall in only one field, i.e., the granite. Likewise, the $Na_2O/Al_2O_3$ vs. $K_2O/Al_2O_3$ plots of the PG also fall in the same field in which the plots of the PRG, GG, and FGGG lie. Compared to the GG (1005 Ma), the pink granite is distinctly 200 Ma younger as revealed by its isochron age of 815±47 Ma. Furthermore, the initial $^{87}Sr/^{86}Sr$ ratio of 0.7538 of the PG is also higher than that of the GG (0.7047).

## Bastar Cratonic Terrain (BCT)

Concentrations of REE are known in granites and associated soils of Paliam (Ramachar et al. 1983), Darba (Singh, 2007), Metapal-Kawadgaon (Singh and Singh, 2001) and a few other areas (Singh, 2002; Singh and Shivkumar, 2009) of Bastar Cratonic Terrain (BCT). Paliam and Darba granites in the southeastern part, and Metapal granite in the northwestern part of BCT are intrusive into Bengpal Group of rocks, with xenoliths of metasedimentary and metabasic rocks, flow banding near the margins and development of an envelope of andalusite schist/gneiss due to contact metamorphism.

## Table 7. Chemical compositions of REE-bearing granites from Bastar Cratonic Terrain (BCT), central India

| Oxide (Wt.%) | Paliam granite | | | | Darba granite | | Metapal granite | |
|---|---|---|---|---|---|---|---|---|
| | Non-greisenised | | Greisenised | | | | | |
| | Range (n = 5) | Average | Range (n = 4) | Average | Range (n = 5) | Average | Range (n = 5) | Average |
| $SiO_2$ | 64.78-74.81 | 71.49 | 71.22-76.40 | 73.74 | 71.93-75.15 | 73.88 | 73.74-77.43 | 75.01 |
| $TiO_2$ | 0.19-0.30 | 0.26 | 0.06-0.27 | 0.17 | 0.29-0.48 | 0.39 | 0.05-0.27 | 0.16 |
| $Al_2O_3$ | 13.22-18.42 | 14.78 | 13.69-13.90 | 13.67 | 12.30-13.69 | 12.71 | 12.14-13.87 | 12.83 |
| $Fe_2O_3$ | 0/04-0.94 | 0.45 | 0.43-0.70 | 0.13 | 0.26-0.78 | 0.59 | 0.10-0.05 | 0.27 |
| FeO | 1.30-2.40 | 1.85 | 0.30-1.51 | 1.10 | 0.94-2.34 | 1.66 | 0.79-2.01 | 1.42 |
| MnO | 0.02-0.06 | 0.04 | 0.01-0.05 | 0.04 | 0.02-0.04 | 0.02 | 0.01-0.05 | 0.03 |
| MgO | 0.02-0.93 | 0.35 | 0.22-0.52 | 0.42 | 0.09-0.44 | 0.34 | 0.09-0.41 | 0.23 |
| CaO | 0.12-1.20 | 0.69 | 0.12-0.60 | 0.37 | 0.27-0.82 | 0.65 | 0.50-1.16 | 0.73 |
| $Na_2O$ | 2.50-4.88 | 3.52 | 1.56-5.06 | 2.85 | 2.50-3.25 | 2.88 | 2.77-3.87 | 3.38 |
| $K_2O$ | 3.38-5.25 | 4.47 | 3.70-6.38 | 5.24 | 5.13-6.75 | 5.78 | 3.19-6.24 | 4.90 |
| $P_2O_5$ | 0.22-0.35 | 0.26 | 0.16-0.30 | 0.22 | 0.04-0.35 | 0.11 | 0.06-0.09 | 0.08 |
| $H_2O^+$ | 0.60-1.80 | 1.48 | 1.00-1.70 | 1.35 | 0.14-2.00 | 0.59 | 0.38-0.69 | 0.53 |
| $H_2O^-$ | 0.20-0.60 | 0.34 | 0.20-0.70 | 0.50 | 0.16-0.70 | 0.21 | 0.12-0.28 | 0.17 |
| Total | 99.55-100.51 | 99.98 | 99.96-100.37 | 100.20 | 99.59-100.27 | 99.81 | 99.51-99.96 | 99.74 |
| A/CNK | 1.15-1.35 | 1.25 | 0.89-1.61 | 1.31 | 0.95-1.28 | 1.06 | 1.01-1.10 | 1.05 |
| Normative corundum | 1.94-4.28 | 3.02 | 1.39-5.41 | 3.48 | 0.00-3.57 | 0.90 | 0.00-3.60 | 0.92 |
| Be (ppm) | ND | | ND | | ND | | 10-21 | 14 |
| Rb | 234-454 | 329 | 475-721 | 601 | 346-478 | 425 | 246-377 | 293 |
| Sr | 78-181 | 121 | 24-78 | 52 | 48-109 | 72 | 33-119 | 75 |
| Ba | 323-954 | 679 | 355-881 | 589 | 285-663 | 462 | 98-685 | 456 |
| Y | 21-45 | 36 | 17-53 | 37 | 28-80 | 51 | 9-27 | 18 |
| Zr | 164-379 | 271 | 71-198 | 163 | 198-313 | 247 | 33-194 | 123 |
| Nb | 17-45 | 30 | 35-52 | 41 | 17-38 | 31 | 26-33 | 30 |
| Sn | 18-36 | 27 | 40-216 | 90 | 13-23 | 18 | 33-54 | 43 |
| Ta | 2.4-7.3 | 4.48 | 1.2-7.8 | 3.7 | ND | ND | ND | ND |
| U | 9-12 | 11 | 7-16 | 11 | 13-28 | 22 | 9-18 | 15 |
| Th | 60-92 | 79 | 48-70 | 59 | 51-117 | 87 | ND | ND |
| Rb/Sr | 1.29-5.82 | 3.10 | 6.37-29.58 | 15.45 | 4.39-9.37 | 6.32 | 2.15-7.45 | 4.48 |
| K/Rb | 114-163 | 128 | 43-107 | 77 | 89-138 | 114 | 108-177 | 133 |

Mineralogically, Paliam granite is two-mica-type, whereas, Darba and Metapal ones are biotite granites (Ramesh Babu, 1999). Quartz, microcline, microcline perthite, and sodic plagioclase ($An_{5-20}$) are the major minerals, whereas, biotite and muscovite form minor minerals. Accessory minerals are apatite, zircon, sphene, allanite, topaz, fluorite, and opaques. The granites vary from monzo-granite to syeno-granite (cf. Streckiesen, 1976). Paliam granite is characterized by greisenised marginal zones (eastern margin) with muscovite reaching upto 13.5 volume%. Presence of fluorite, topaz, and cassiterite is notable, whereas, monazite and/or xenotime constitute REE-bearing phases.

Analytical data (Table 7) show that these three granitic bodies are comparable in their chemical composition. The granites are peraluminous with A/CNK value of > 1.0 and average (av.) normative corundum of 0.90 – 3.02. Compared to normal granites, these are higher in

SiO$_2$ (av.71.49 to 75.01%) and K$_2$O (av. 4.47 to 5.78%) and lower in TiO$_2$ (av. 0.16 to 0.39%), MgO (av. 0.23 to 0.35%), and CaO (av. 0.65 to 0.73%). Among the three, Paliam granite is more peraluminous (A/CNK 1.15 – 1.35, av. 1.25; normative corundum 1.94 - 4.28, av. 3.02), lower in SiO$_2$ (64.98 - 74.81%, av. 71.49%) and K$_2$O (3.38 - 5.25%, av. 4.47%). Metapal granite is more silicic (73.74 -77.43%, av. 75.01% SiO$_2$) and Darba granite is more potassic (5.13 – 6.75%, av. 5.78% K$_2$O). In K$_2$O – Na$_2$O – CaO diagram, the granites plot in the fields for adamellite and granite. These are enriched in Rb (av. 293 – 425 ppm), Y (av. 36 and 51 ppm in Paliam and Darba), Nb (av. 30-31 ppm), Sn (av. 18-43 ppm), U (av. 11-22 ppm), Th (av.79-87ppm), Be (10-21, av. 14 ppm in the case of Metapal granite) and also higher Rb/Sr (av. 3.10 – 6.32) and lower K/Rb (av. 114 – 133). Compared to the Paliam and Metapal granites, Darba granite is marked by higher contents of Rb (346 – 478 ppm, av. 425 ppm), Y (28-80 ppm, av. 51 ppm), U (13-28 ppm, av. 22 ppm), and Th (51-117 ppm, av. 87 ppm), lower in Sr (48-109 ppm, av. 72 ppm), and K/Rb (89-138, av.114) with higher Rb/Sr (4.39-9.37, av. 6.32). Metapal granite contains the highest Sn (33-54, av. 43 ppm). In the Rb-Sr-Ba ternary diagram, majority of the samples of these granites plot in the field for strongly differentiated granite with a few in the field for normal granite. A syn-collisional tectonic setting is indicated by Rb vs. Y + Nb plots of these granites. Greisenised, muscovite-rich marginal zones of Paliam granite typically containing fluorite and topaz, compared to its non-greisenised interior, are distinctly marked by higher amounts of Rb (475-721 ppm, av. 601 ppm), Nb (35-52 ppm, av. 41 ppm), and Sn (40-216 ppm, av. 90 ppm) as well as Rb/Sr (6.37-29.58, av. 15.45) and lower K/Rb (43-107, av. 77) denoting that the Paliam granite is subjected to an intense late alkali and F-metasomatism (Ramesh Babu et al. 1984).

Total REE content in Paliam granite (Table 8), in general, is high, with relative enrichment of LREE over HREE [(Ce/Yb)$_{cn}$ of 1.74-21.27]. The chondrite-normalised REE patterns show a gentler downward slope for LREE and a more or less flat pattern for HREE, with pronounced negative Eu-anomaly (Eu/Eu* = 0.23-0.38; mean 0.31). Higher (Ce/Yb)$_{Cn}$ values (4.06-21.27) in the non-greisenised granite indicate a higher degree of fractionation. Compared to non-greisenised granite, the greisenised zones are marked by relatively lower content of total REE, LREE, HREE and (Ce/Yb)$_{Cn}$, and higher degree of negative Eu-anomaly (Eu/Eu* = 0.23-0.35, av. 0.27).

Table 8. REE data (ppm) on the REE-bearing Paliam granites from Bastar Cratonic Terrain, central India

| Element | 1 | 2 | 3 | 4 | 5 | 6 | 7 | 8 |
|---|---|---|---|---|---|---|---|---|
| La | 94 | 118 | 130 | 132 | 141 | 62 | 50 | 31 |
| Ce | 152 | 187 | 213 | 205 | 255 | 99 | 89 | 68 |
| Sn | 8.2 | 17 | 14 | 15.2 | 18.4 | 6.6 | 10.5 | 5.8 |
| Eu | 0.85 | 1.7 | 0.8 | 1.5 | 1.9 | 0.5 | 101 | 0.5 |
| Ta | 1.2 | 1.75 | 1.5 | 1.85 | 2.5 | 1.1 | 1.5 | 1.45 |
| Yb | 2.2 | 7.7 | 2.5 | 6.6 | 15.2 | 3.4 | 5.0 | 9.5 |
| Lu | 0.32 | 1.1 | 0.35 | 0.95 | 2.4 | 0.45 | 0.7 | 1.45 |
| Total | 258.77 | 334.25 | 362.15 | 363.10 | 436.4 | 173.05 | 157.8 | 117.7 |
| (Ce/Yb)$_{Cn}$ | 16.7 | 5.86 | 21.27 | 7.5 | 4.06 | 7.27 | 4.26 | 1.74 |
| Eu/Eu* | 0.30 | 0.38 | 0.24 | 0.36 | 0.36 | 0.23 | 0.35 | 0.23 |

## ORIGIN OF REE-BEARING GRANITES

### Chhotanagpur Granite Gneiss Complex (CGGC) Terrain

*Petrography and Modal Composition:* A critical evaluation of the petromineralogical and modal compositions of the PRG, GG, FGGG, and PG reveals that the mineralogy of the texturally-different types of granitoids of the Raikera-Kunkuri area is more or less similar. Furthermore, triangular plots of modal Q-A-P of the different types of granitoids also fall in the same field, that is, the granite. However, the abundances of accessory minerals show noticeable variations (Tables 1, 2, 4, and 5). Accordingly, based on the IUGS recommended parameters (Streckeisen, 1976), the granitoids could be classified as: (i) PRG: hornblende-biotite granite and biotite granite, (ii) GG: muscovite-biotite granite and muscovite granite, (iii) FGGG: biotite granite, and (iv) PG: muscovite-biotite granite.

Despite the similarities, there is also a significant difference between the various types of the granitoids due to certain accessory minerals and petrographic features. For example, the accessory mineral hornblende, highest content of biotite (av. 12.1 volume %), presence of magnetite, absence of muscovite, and only scarce grains of monazite, as compared to the other types, characterise the porphyritic granite. In contrast, the grey granite distinguishes itself from the others on account of the intermediate abundance of biotite (av. 7.2 volume %), common presence of coarse-grained muscovite, and increased frequency of monazite grains. Relative to these two, the conspicuous features of the FGGG are the highest silica content, modal abundance of quartz (av. 32.8 volume %), and K-feldspar (av. 37.2 volume %), and the least abundance of biotite (av. 5.6 volume %), and negligible muscovite, besides the presence of the heavy rare earth mineral, xenotime. Apart from the pink colour, the PG displays distinctly clouded feldspar and development of sericite resulting from hydrothermal alteration and presence of comparatively higher abundance of xenotime grains. It may, therefore, be concluded that, in addition to textural variations, various types of the granites show notable differences in accessory mineralogy, although the modal abundances of their essential mineral constituents place them in only one group, that is, granite.

The presence of hornblende in some samples (Table 1), magnetite, and sphene in PRG, and the progressive decrease of the dark-coloured mineral, biotite, from the PRG to GG, coupled with the presence of coarse-grained muscovite, including increased abundance of light-coloured heavy minerals, zircon and monazite, in GG, possibly suggest consanguineous nature of the felsic melt from which both PRG and GG formed. Textural and mineralogical variations appear to have been caused due to variable depths of emplacement. As is well-known from the Bowen's reaction series, the formation of hornblende precedes biotite crystallisation in a melt. Although hornblende is not present in all the samples of the PRG (Table 1), wherever present, it shows transition into biotite. Similarly, the accessory ore mineral, magnetite, also forms early when crystallisation of the melt begins. By analogy, therefore, the PRG, hosting hornblende and magnetite, could be considered to represent an early differentiate of the consanguineous felsic melt. The fact that the crystallisation of the PRG took place under elevated pressure and $H_2O$ content is supported by the presence of dominant mafic phases, hornblende (grading from brown to green) and biotite together (Winter, 2001, p. 344). The coarse-grained porpyritic texture of the PRG could be due to its emplacement and crystallisation at deeper level, where prevalence of elevated pressure is also expected. The appearance of hornblende, a preponderance of biotite, and only scarce grains of monazite, could be due to less differentiation of the melt. In contrast, after formation of the early differentiate, a comparatively more evolved melt moved upward where crystallisation of the medium-grained GG took place at relatively shallower level, which was more fractionated and characterised by the absence of hornblende and magnetite, reduced content of biotite, common presence of coarse-grained muscovite, and the increased abundance of monazite and apatite grains. Predominance of primary muscovite crystals also indicates that the GG formed during pluton cooling (Winter, 2001, p. 344).

In contrast, the FGGG, perhaps, represents a separate episode of more evolved younger granite magmatism of limited areal extent at a relatively shallower level that facilitated development of fine-grained texture. The presence of the heavy rare-earth mineral, xenotime, testifies to its being a separate episode of granite magmatism of a more evolved nature, as heavy REEs always concentrate in more fractionated melts (Mariano, 1989). If the xenotime-bearing FGGG had genetic links with the melt from which PRG and GG were derived, xenotime would have been present in them also. It is likely that the younger granite magmatism (responsible for FGGG) and the attendant hydrothermal activity were also responsible for the metasomatism and the development of pink granite in the CGGC terrain. Sericitisation of feldspar and chloritisation of biotite also support hydrothermal alteration of the PG (Thuy Nguyen et al. 2004). Since hydrotherms derived from the melt of the FGGG were more fractionated, concentration of xenotime is higher in the PG (Singh and Singh, 1996). The limited areal extension of PG and its proximity to FGGG supports this interpretation. The PG and GG may represent the same initial episode of granite magmatism based on textural similarity of grain size; and the pink colour in PG may be attributed to flushing of post-magmatic fluids.

*Petrochemistry:* Synthesis of whole-rock data suggests that, similar to the modal compositions, the chemical compositions of the granitoids are also, by and large, similar. Chemically, like the modal classification, all the granitoids could also be classified as granite (ss). Furthermore, all of them are peraluminous and record $K_2O/Na_2O$ ratios more than 1.0. Interestingly, they also seem to have been derived from the anatexis of common crustal materials of similar compositions, a fact revealed by their $Na_2O/Al_2O_3$ vs. $K_2O/Al_2O_3$ plots of all the types of granitoids in the same field, i.e., sedimentary/metasedimentary. However, on closer scrutiny of the compositional data, like the mineralogical characteristics, a few notable chemical variations are also apparent. The PRG shows a distinct chemical transition from adamellite to granite. Furthermore, not only has it the lowest $SiO_2$ (av. 68.17%) content, but also the lowest $K_2O/Na_2O$ (av. 1.43), $Fe_2O_3/FeO$ (av. 0.76), and $Fe_2O_3/MgO$ (av. 2.16)) ratios. In contrast, the GG is, chemically, truly restricted to granite composition. Also, compared to the PRG, it has more $SiO_2$ (av.72.17%) and distinctly higher $K_2O/Na_2O$ (av. 2.03), $Fe_2O_3/FeO$ (av. 0.96), and $Fe_2O_3/MgO$ (av.3.31) ratios. The petrochemical characteristics of both the types clearly demonstrate that the PRG is chemically less evolved compared to the GG. Indeed, it is due to the more chemically evolved nature of the GG that its silica content and $K_2O/Na_2O$ and $Fe_2O_3/MgO$ ratios are higher, relative to the PRG. Interestingly, the $Fe_2O_3/FeO$ ratio of less than 1 is known to indicate emplacement of a melt at deeper level, whereas, a value of 1 and above suggests depth of emplacement at shallower levels (Dikshitulu and Dhana Raju, 1997). Significantly, therefore, notably low $Fe_2O_3/FeO$ (av. 76) ratio of the PRG may be taken to indicate its emplacement at deeper levels. As against this, the $Fe_2O_3/FeO$ ratio of the GG near unity (av. 0.96) could reveal its emplacement at relatively shallower depths. Such variable $Fe_2O_3/FeO$ ratios with respect to depths of emplacement are likely to be brought about by the oxygen fugacity-controlled changing redox conditions attending the crystallisation history of the rising melt at different depths. These petrochemical variations, including ratios, are in conformity with the observed mineralogical features of the PRG and GG.

As against the PRG and GG, the highest amount of $SiO_2$ (av. 73.35%) in the FGGG is noteworthy. In addition, the FGGG has intermediate ratios of $K_2O/Na_2O$ (av. 1.71), $Fe_2O_3/FeO$ (av. 0.85), and $Fe_2O_3/MgO$ (av. 2.59). The highest amount of silica in the FGGG could be because of involvement of more silicic crustal material in its genesis. Significantly, the PG has $SiO_2$ (72.08%) almost similar to that of the GG (av. 72.17%). Also, the $K_2O/Na_2O$ ratio of the PG (2.08) and the GG (2.03) are close to each other. On the other hand, $Fe_2O_3/FeO$ (0.88), and $Fe_2O_3/MgO$ (2.58) ratios of the PG are closer to such ratios (0.85 and 2.59, respectively) of the FGGG. The $^{87}Sr/^{86}Sr$ isotopic ratio (0.7538) of the PG is distinctly higher than this ratio in the GG (0.7047). Similar to the signature of grain size, based on similarity of $SiO_2$ contents and $K_2O/Na_2O$ ratios also the grey granite (GG) and pink granite (PG) represent a single initial episode of granite magmatism. However, the similarities in the values of $Fe_2O_3/FeO$ and $Fe_2O_3/MgO$ ratios of the PG and FGGG may, perhaps, be linked with the metasomatism of the earlier formed GG, consequent to the emplacement of the FGG. Nevertheless, it appears that the younger episode of granitic activity which affected PG did not cause any change in the bulk composition of PG from the GG, except in the heavy REE mineralisation. Accordingly, the metasomatism may also be responsible for the higher isotopic evolution and more isotopic ratio of the PG.

*Petrogenesis:* The CGGC terrain is known to have undergone polyphase metamorphism (Mahadevan, 2002 and 2008; Patel et al., 2007; Maji et al., 2007; Ramakrishnan and Vaidyanadhan, 2008). The metamorphism and attendant metasomatism undergone by the metamorphites of the CGGC region attained a temperature of 640-800°C and a pressure of 3-6 kb (Bhattacharyya, 1988; Sarkar, 1988; Ghose, 1983 and 1992). Also, from the generation of felsic melts in syn-tectonic environments, the 1005 Ma grey granite (including porphyritic granite by analogy) is coeval to the $F_3$ deformation and $M_3$ metamorphism during 1000-850 Ma that is reported to be responsible for the formation of massive granites (Sarkar, 1982 and 1988). This event post-dates the $F_1$ deformation and $M_1$ metamorphism dated at ~2000-1600 Ma, and evolution of granite gneisses at ~1500-1170 Ma, and the deformation cycles $F_1$ and $F_2$, reported from the central and eastern parts of the CGGC (Sarkar, 1982 and 1988). Accordingly, it is likely that the generation of granitic melt took place in response to $F_3$ deformation and $M_3$ metamorphism. After generation of the melt, its initial emplacement and crystallisation took place at deeper levels, where an early differentiate in the form of porphyritic granite came into being, with the development of coarse-grained porphyritic texture, and retention of some relicts of hornblende. Later, a relatively chemically more evolved melt moved to a shallower level, where crystallisation of the medium-grained GG took place, accompanied by the development of coarse-grained muscovite at the expense of biotite. This interpretation is supported by the fact that the $Fe_2O_3/FeO$ ratios of the PRG are the lowest (av. 0.76), whereas, it is near unity in the case of GG (av. 0.96). In addition, the higher $Fe_2O_3/MgO$ ratio of the GG also supports its chemically more fractionated nature, as compared to the PRG. Plots of $Na_2O/Al_2O_3$ vs. $K_2O/Al_2O_3$ and peraluminous nature and also reveal that the PRG and GG are the products of a common melt derived from crustal material and from the metamorphism and anatexis of a metasedimentary suite. The presence of older metasedimentary rocks in the CGGC terrain also supports this interpretation.

The distinctly fine-grained nature and limited areal extent of the FGGG makes it possibly a phase different from the PRG and GG. Accordingly, it may be taken to represent a localised granitic activity, unrelated to the former. In view of the distinct isochron age of the pink granite (815 Ma), with signatures of metasomatism in PG and the proximity of PG to FGGG, it appears that the FGGG was, perhaps, responsible for the genesis of the PG. Because of the small-scale nature of the FGGG intrusion, the extent of metasomatism was also restricted; hence, the areal extension of the PG is also limited. The isochron of the PG also records signatures of the metasomatic/metamorphic disturbances (Singh and Krishna, 2009). By analogy, therefore, the 815 Ma age of the PG may also be taken to represent the age of the FGGG. Accordingly, the genesis of the FGGG melt might be linked with the late metamorphic processes, which are also considered to be responsible for the resetting of the Rb-Sr isotopic system and more SrI of 0.7538 of the PG (Singh and Krishna, 2009). Field observations such as the limited areal extent of the pink granites, and their transitional and irregular gradations into other variants, seem to support the fact that the pink granites are the product of metasomatic activities (Singh and Singh, 1996). In addition, from various parts of the CGGC terrain (Mahadevan, 1992; Singh and Rai, 2008) also, pink granites of metasomatic origin are known. Therefore, the ~815Ma age of the investigated pink granites might represent the time of metasomatism related to the second phase of granitic activity in the area.

*Phases of REE- and Y-mineralisation:* The available isochron ages of REE-bearing granites from the Kunkuri-Raikera area of the CGGC terrain indicate two phases of the granitic activities. The first phase is represented by the emplacement of grey (and, by analogy, phorphyritic) granite at 1005 Ma and the second phase of pink (and, possibly, also fine-grained grey) granite at 815 Ma (Singh and Krishna, 2009). The ubiquitous presence of monazite in all types of granitoids of the Kunkuri area is conspicuous (Singh and Singh, 1996). This distribution pattern of monazite, therefore, reveals that light rare earth element (LREE) mineralisation in the form of monazite took place during both the phases of the granitic activities known from the Raikera-Kunkuri area. The first phase of the LREE mineralisation was associated with the emplacement of the large-scale granitic bodies in the form of PRG-GG at ~1005 Ma, whereas, the second phase was linked with the later phase of granitic activity at 815 Ma, represented by the FGGG and associated pink granite. In contrast, the heavy rare earth element (HREE) mineralisation in the form of xenotime, hosted in the FGGG and PG, took place at ~815 Ma. The one and only phase of xenotime mineralisation was accompanied by the second phase of the LREE, monazite, mineralisation. However, robust ages of zircon, xenotime, and monazite, hosted in different granitoids, are needed in constraining the REE metallogeny in time and space in CGGC terrain of central India.

## Bastar Cratonic Terrain

Early intrusive syntectonic granitic activity, involving anatexis and related migmatisation of pre-existing rocks forming gneissic granites and migmatites, took place in the Neoarchaean, for which Rb-Sr, whole-rock, isochron ages range from 2528 to 2659 Ma (Sarkar et al. 1990 a and b). The gneisses around Sukma represent a typical tonalitetrondjhemite-granodiorite suite (Karkare and Srivastava, 1985). Subsequently, the area witnessed the main granite tectonism in the Palaeoproterozoic, which culminated in the emplacement of pegmatites (host rocks for rare-metal and also rare-earth mineralization) together with some quartz veins in the weak zones in the metabasic and metasedimentary rocks, and the granites. The available whole-rock, Rb-Sr ages of granitic rocks from different parts of Bastar craton (Paliam, Darba: 2275 ± 80 Ma; Burugudam: 2237 ± 70 Ma; Pujariguda: 2110 ± 47 Ma; Cholanguda: 2301 ± 53 Ma; Pandey et al. 1989) suggest that remobilization of sialic crust, resulting in the emplacements of granitic rocks from an anatectic melt, took place during the period ca 2300-2100 Ma. The presence of highly altered andalusite tabloids (1-4 cm) in the schists bordering the granites points to the contact-metamorphic effect, and the emplacement of granites at shallow depth in the crust (Lamba and Agarkar, 1988; Mishra et al. 1988). However, younger K-Ar age of 1345 ± 22 Ma for biotite from granitic rocks exposed in the vicinity of village Paliam (Anonymous, 1978) may be due to the effect of loss of Ar during some late geological event. Quartz veins, traversing all the older rocks including pegmatites, perhaps mark the last phase of acid magmatism in the area. The metasedimentary and metabasic rocks, granites, and the pegmatites have been intruded by dolerite dykes.

In and around Kawadgaon also, where pegmatites are currently being mined for recovering rare-metal and associated rare-earth minerals, the rocks of Bengpal Group are intruded by Palaeoproterozoic rare-metal-bearing granites-pegmatites (Singh, 1999). Wherever observable, the contact between the felsic intrusive and the Bengpals is either sharply defined, or marked by silicification, development of greisen, or concealed. The granites are fine- to medium-grained, mesocratic to leucocratic, tending to two-mica (biotite and muscovite) type. The intrusive Palaeoproterozoic granites show notable transition to pegmatitic granite, and also to simple and zoned pegmatites. Such transitions suggest that the pegmatites, with rare-metal concentrations, are differentiates of their host Palaeoproterozoic granites (Singh, 1998). Furthermore, the granites show vertical pegmatite zonation in their apical parts. The pegmatites at lower-middle level contain beryllium and tantalo-niobate mineralization and those at higher level carry tin concentration, with grains that can be handpicked of monazite and xenotime in some of the pegmatites at lower-middle level. The granites have yielded Rb-Sr, whole-rock, isochron age of 2497±152 Ma (Singh and Chabria, 1999). The oldest age of the muscovite (2329 Ma) from the co-genetic pegmatites of Kawadgaon nearly overlaps within 2 sigma error of the parent fertile granites. Some of the muscovite ages (1848-2156 Ma) from the co-genetic pegmatites, which were derived shortly after the emplacement of their co-genetic granites, also indicate tectonomagmatic ages after pegmatite injections (Singh and Chabria, 2002). The granitoids of the Kawadgaon appear to be older than those from Paliam and Darba granites in the southeastern part of the Bastar craton (2308 Ma; Bandyopadhyay et al. 1990; Ramesh Babu et al. 1993).

# RARE-EARTH-ELEMENT MINERAL RESOURCES

## Chhotanagpur Granite Gneiss Complex (CGGC) Terrain

*Rare-Earth-Element and Yttrium:* Large resources of light rare-earth-element (LREEs) are contained in beach placers of eastern and southern sea coasts of India. As against this, heavy rare-earth-element (HREEs) and Y resources are known only from inland placers. In this context, CGGC terrain forms potential belt not only for HREEs and Y, but also for associated LREEs minerals (and limited Nb-Ta mineral). Rare-earth-element and yttrium (REE and Y) mineral resources are genetically connected to polyphase 'S'-type granites of CGGC. Fine-grained grey granite (FGGG) and pink granite (PG) in and around Kunkuri and other areas of the CGGC terrain are identified to be the main sources of heavy REE and Y minerals (Singh, 1990). On the other hand, apart from FGGG and PG, grey and porphyritic granites also contribute light REE mineral (Singh, 1992). Riverine placers along streams draining such granites contain economic concentrations of REE and Y minerals (Singh, 1990; Singh and Singh, 1996). The Y and HREE phase is represented by xenotime, whereas, monazite represents the LREE phase. Workable concentrations along stream placers are located in northeastern parts (Kunkuri-Jashpur region) of Chhattisgarh, central India.

In most REE-bearing alluvial deposits of the river, well-defined, lateral and vertical zonation of radioactivity is noticed. In the first case, the radioactivity is maximum in river banks, which gradually decreases laterally. In the second case, the upper surface of the deposit records the highest radioactivity, which gradually decreases vertically downward and becomes negligible after about 1 m depth. Only in isolated cases, it continues up to a depth of 2.5 m. Such a behavior of radioactivity is due to lateral and vertical zonations in the concentrations of heavy minerals.

The mineralogical data indicate that the gross mineralogy of different alluvial deposits is the same. The heavy minerals include magnetite, hematite, and ilmenite together with some garnet, epidote, and fluorapatite. REE and Y-bearing minerals are monazite and xenotime, besides zircon and some rutile. Interestingly, monazite is ubiquitously present in alluvial deposits of almost all the rivers of the lower orders (up to $5^{th}$ order), whereas, xenotime is present only in a few river deposits. However, the rivers of the higher, viz., $6^{th}$ and $7^{th}$, orders neither contain monazite, nor xenotime (Singh and Rai, 1992). Data on polymineral table concentrates, containing monazite, xenotime, zircon, ilmentite, and some rutile, indicate 1.18 to 3.08% $Y_2O_3$ (av. = 2.34%), 0.127 to 0.300% $Er_2O_3$ (av. = 0.238), and 0.0106 to 0.0286% $Eu_2O_3$ (av. = 0.02%), with low values of standard deviations and coefficients of variation. These indicate definite decrease in assay values from $Y_2O_3$ to $Er_2O_3$ to $Eu_2O_3$ in the mutual ratio of 10:1:0.1. Xenotime contains upto 42.30% $Y_2O_3$ and is also radioactive (Table 9).

Significantly, numerous occurrences of rare-earth-bearing riverine placers, containing xenotime and monazite, are also known from many streams of the Kanhar (Ramesh Babu et al. 1998) and Mahan (Rai and Banerjee, 1995) river basins. Panned concentrates from Kanhar river basin revealed 0.42 to 4.83% $Y_2O_3$ and 0.50 to 8.77% $Ce_2O_3$, whereas, concentrates from Mahan river basin recorded 0.20 to 3.0% $Y_2O_3$ and 1.59 to 23.24% $Ce_2O_3$. Initial studies have indicated substantial upgradation of yttrium (upto 20.92% $Y_2O_3$) and cerium (20.13% $Ce_2O_3$) contents in induced roll magnetic separates at different Kilo Gauss (Rai and Banerjee, 1995), suggesting encouraging upgradation potential of these REE-bearing placer deposits.

A comparison of heavy mineral concentrates from selected streams in Kanhar, Mahan, and IB (Siri) river basins is shown in Table 10. Table concentrates from Bairamua and Sarsoti nalas in Kanhar river basin contain 6.0 to 7.4% xenotime (Table 10). This is higher than those of Joj and Garahari nalas (5.1% and 2.7%, respectively) in Mahan river basin and also Siri river (3.5%). Compared to Siri river, the concentrates from the streams of Kanhar river basin are poor in ilmenite, rich in garnet, and monazite contents (Table 10).

**Table 9. Chemical composition of xenotime from Siri river placers, CGGC terrain, central India**

| Oxide | RBX | YBX | YTX |
|---|---|---|---|
| $La_2O_3$ | 0.40 | 0.45 | 0.54 |
| $CeO_2$ | 0.75 | 0.71 | 0.95 |
| $Pr_6O_{11}$ | 0.12 | 0.11 | 0.11 |
| $Nd_2O_3$ | 0.53 | 0.50 | 0.60 |
| $Sm_2O_3$ | 0.51 | 0.51 | 0.51 |
| $Eu_2O_3$ | 0.025 | 0.031 | 0.031 |
| $Gd_2O_3$ | 1.65 | 1.54 | 1.50 |
| $Tb_4O_7$ | 0.52 | 0.47 | 0.45 |
| $Dy_2O_3$ | 4.68 | 4.60 | 4.46 |
| $Ho_2O_3$ | 1.12 | 1.12 | 1.10 |
| $Er_2O_3$ | 4.06 | 4.10 | 4.12 |
| $Tm_2O_3$ | 0.64 | 0.67 | 0.68 |
| $Yb_2O_3$ | 4.00 | 4.55 | 4.60 |
| $Lu_2O_3$ | 0.59 | 0.69 | 0.68 |
| $Y_2O_3$ | 42.30 | 41.00 | 40.10 |
| $ThO_2$ | 1.10 | 1.07 | 0.96 |
| $U_3O_8$ | 0.82 | 1.11 | 1.17 |
| $P_2O_5$ | 30.00 | 32.00 | 31.70 |
| $SiO_2$ | 1.20 | 1.00 | 1.00 |
| $Fe_2O_3$ | 1.30 | 0.90 | 0.80 |
| $Al_2O_3$ | 0.40 | 0.40 | 0.20 |
| CaO | 0.35 | 0.38 | 0.39 |
| PbO | 0.16 | 0.15 | 0.19 |
| $Zr_2O_3$ | 0.02 | 0.02 | 0.02 |
| Total | 97.23 | 98.06 | 96.84 |

RBX = Reddish brown. YBX = Yellowish brown. YTX = Pale yellow transparent.

**Table 10. Comparison of heavy minerals concentrates (all data in weight percent) among major riverine placers of CGGC terrain, central India**

| Sl. No. | Minerals | Kanhar river basin | | Mahan river basin | | Ib river basin |
|---|---|---|---|---|---|---|
| | | Bairamua nala | Sarsoti nala | Jaj nala | Garhari nala | Siri river |
| 1. | Xenotime | 6.0 | 7.4 | 5.1 | 2.7 | 3.5 |
| 2. | Monazite | 31.0 | 54.4 | 31.7 | 22.3 | 38.0 |
| 3. | Ilmenite | 43.0 | 9.0 | 36.6 | 42.0 | 45.0 |
| 4. | Garnet | 11.0 | 24.8 | 18.0 | 22.6 | 5.0 |
| 5. | Zircon | 6.0 | 4.0 | 3.6 | 2.4 | 4.0 |
| 6. | Magnetite | 3.0 | 0.4 | - | 2.0 | 3.0 |
| 7. | Others | - | - | 5.0 | 6.0 | 1.5 |
| Total | | 100.0 | 100.0 | 100.0 | 100.0 | 100.0 |

**Table 11. Chemical compositions (all data in weight percent) of monazites from Bastar pegmatites, central India, along with composition of monazite from Trout Creak Pass Pegmatite (TCPP), Colorado, USA**

| Oxides | Range (n = 6) | Average | Monazite from TCPP |
|---|---|---|---|
| $La_2O_3$ | 7.12 - 9.8 | 8.22 | 8.73 |
| $CeO_2$ | 18.3 - 23.0 | 20.05 | 24.07 |
| $Pr_6O_{11}$ | 1.91 - 2.17 | 1.96 | 1.49 |
| $Nd_2O_3$ | 9.24 - 11 | 9.9 | 11.20 |
| $Sm_2O_3$ | 4.04 - 4.92 | 4.65 | 3.31 |
| $Eu_2O_3$ | 0.02 - 0.05 | 0.03 | 0.49 |
| $Gd_2O_3$ | 2.30 - 3.40 | 2.77 | 2.72 |
| $Tb_4O_7$ | 0.42 - 0.57 | 0.48 | NA |
| $Dy_2O_3$ | 0.65 - 0.90 | 0.76 | 0.49 |
| $Ho_2O_3$ | 0.05 - 0.065 | 0.07 | NA |
| $Er_2O_3$ | 0.05 - 0.075 | 0.075 | NA |
| $Tm_2O_3$ | 0.008 - 0.011 | 0.008 | NA |
| $Yb_2O_3$ | 0.04 - 0.037 | 0.04 | NA |
| $Lu_2O_3$ | 0.002 - 0.010 | 0.005 | NA |
| $Y_2O_3$ | 1.64 - 2.10 | 1.91 | 2.07 |
| $U_3O_8$ | 0.164 - 0.331 | 0.276 | NA |
| $ThO_2$ | 11.9 - 16.2 | 13.66 | 11.33 |
| $SiO_2$ | 1.3 - 5.6 | 2.88 | 3.02 |
| $P_2O_5$ | 24.5 - 25.8 | 24.42 | 29.48 |
| $Fe_2O_3$ | 0.44 - 4.40 | 2.55 | NA |
| MnO | <0.01 - 0.035 | 0.03 | NA |
| $Al_2O_3$ | 0.10 - 1.15 | 0.70 | NA |
| MgO | 0.01 - 0.03 | 0.02 | NA |
| $Na_2O$ | <0.1 - 0.10 | - | NA |
| $K_2O$ | <0.1 - 0.10 | - | NA |
| Pb | 0.66 - 1.48 | 0.85 | 0.90 |

Sarah et al. 1992. NA = Not available.

## Bastar Cratonic Terrain

Rare-Earth-Element and Yttrium: In the Bastar cratonic terrain known resources of REEs are limited. Rare-earth-bearing minerals are represented by monazite and xenotime, which occur in pegmatites and their parent granites. Bigger size and well-developed euhedral crystals of monazite occur in association with xenotime and Nb-Ta minerals in pegmatites. Monazite from the rare-metal-mineralised pegmatites of Bastar is yellowish or reddish brown to brown and earthy. It occurs as fine to coarse fragments, lumps, and crystals. Bigger monazite crystals display flattened [100] or elongated [100] habit, and show development of (100) or (111) faces. Crystal faces are smooth, pitted, or striated. Pure, hand-picked monazite crystals from Bodenar-Kawadgaon pegmatite reveal high absolute values of La, Ce, and P, with appreciably high Nd contents. The monazites from Bastar pegmatites contain unusually high ThO$_2$ (11.9-16.2%, average 14%). Although in monazite ThO$_2$ may be present upto 12%, grading to cheralite (Gaines et al. 1997, p. 723 ), in monazites from Bastar pegmatites absolute content of ThO$_2$ (av. 14%) is still more, and is next to CeO$_2$ (18.3-23.0%, average 20%) content (Table 11). Accordingly, based on these two dominant elements, monazites from Bastar pegmatites belong to a distinct class, 'Thorium-rich Monazite-(Ce)', falling in between monazite and cheralite, although more towards monazite end (Singh et al. 2011). Compositionally, monazites from Bastar pegmatites appear to be akin to monazite from pegmatite of Trout Creek Pass, Colorado, USA (Sarah et al. 1992). Panned concentrates from granitic soils of the Kawadgaon and adjoining area, with occasional xenotime and monazite (Singh and Singh, 2001), revealed low and uneconomic yttrium (0.002 to 0.57% Y) and cerium (0.01to1.43% Ce) contents.

Anomalously high concentrations of yttrium have been shown by granitic soil deposits in and around Karangal Dongri near Darba (Singh, 2007). Hand panned concentrates of soil revealed 0.13 to 2.26% Y$_2$O$_3$ and 0.05 to 3.21% Ce$_2$O$_3$, with notable contents of Nb, Ta, and Sn. Preliminary beneficiation studies have shown the possibilities of upgrading yttrium and cerium contents in soil concentrates. In both coarse (+100 #) and fine (-100 #) fractions, the highest values of yttrium are contained in magnetic at 0.45 ampere (3.59-7.69% Y$_2$O$_3$) and that of cerium in magnetic at 0.60 ampere (3.0-8.06% Ce$_2$O$_3$; in -100#, highest value is 4.32% Ce$_2$O$_3$ in magnetic fraction at 1.0 ampere). Some xenotime and monazite occurrences are also known from Jamair-Paliam area of Bastar cratonic terrain (Ramachar et al. 1983).

Granitic soil / riverine placers, due further west of Dantewada, that is, western side of N-S trending Bailadila hill ranges, in the environs of the Mari river around Midtul (Survey of India toposheet no. 65F/1; 18$^0$52'30": 81$^0$09'55") record poor abundances of rare-earth-elements. Yttrium content ranges from 0.006 to 0.670 (av. 0.092) % Y$_2$O$_3$ and cerium from 0.019 to 3.861 (av. 0.834) % Ce$_2$O$_3$ (Singh and Shiv Kumar, 2009). The data indicate a preponderance of cerium over yttrium. Further, in and around Amraoti (Survey of India toposheet no. 65E/14) also, panned concentrates of granitic soils record similar low contents (Singh, 2002) of Y (0.022 to 0.992% Y$_2$O$_3$) and Ce (<0.010 to 2.74 Ce$_2$O$_3$), including Nb (0.016 to 0.208% Nb$_2$O$_5$), Ta (<0.010 to 0.325% Ta$_2$O$_5$), and Sn (<0.01 to 1.481% SnO$_2$).

## CONCLUSION

The rare-earth-element-bearing granitoids of the Precambrian Chhotanagpur Granite Gneiss Complex (CGGC) terrain are texturally three types (i) coarse-grained porphyritic granitoid (PRG), (ii) medium-grained (a) grey granitoid (GG) and (b) pink granitoid (PG), and (iii) fine-grained grey granitoid (FGGG). Based on the IUGS recommended parameters, the granitoids can be classified as (i) PRG: hornblende-biotite granite and biotite granite, (ii) GG: muscovite-biotite granite and muscovite granite, and (iii) FGGG and PG: biotite granite and muscovite-biotite granite, respectively. Chemically, like the classification based on modes, all the granitoids can also be classified as granite (ss).

The Na$_2$O/Al$_2$O$_3$ vs. K$_2$O/Al$_2$O$_3$ plots reveal that the PRG, GG, FGGG, and PG are the products of the melts derived from crustal material by metamorphism and anatexis of metasedimentary suites. After generation of the melt, its initial emplacement and crystallisation took place at deeper level, where an early differentiate as porphyritic granite formed. Later, a relatively more chemically evolved melt moved to a shallower level, where crystallisation of the medium-grained GG took place. The distinctly fine-grained nature and limited areal extent of the FGGG make it possibly a phase different from the PRG-GG phase. As against 1005 Ma age of the GG, the distinctly younger isochron age of the pink granite (815 Ma), with signatures of metasomatism in PG and the proximity of PG to FGGG, it appears that the FGGG was, perhaps, responsible for the genesis of the PG. Accordingly, the genesis of the FGGG melt and PG might be linked with the younger metamorphic processes at ~815 Ma. The first phase of the LREE mineralisation was associated with the emplacement of the large-scale granitic bodies in the form of PRG-GG at ~1005 Ma, whereas, the second phase was linked with granitic activity at ~815 Ma, represented by the FGGG and pink granite. In contrast, the heavy rare-earth-element (HREE) mineralisation in the form of xenotime, hosted in the FGGG and PG, took place at ~815 Ma, which was accompanied by a second phase of the LREE mineralisation.

In the Bastar Cratonic Terrain (BCT), limited concentrations of REE are known in granites and pegmatites and associated soils in various parts. Mineralogically, REE-bearing felsic bodies are biotite and two-mica-type monzo-granite to syeno-granite. Monazite and/or xenotime constitute REE-bearing phases. Geochemical data show that these granitic bodies are comparable in their chemical compositions. They are peraluminous with normative corundum. A syn-collisional tectonic setting is indicated in Rb vs Y + Nb plots for these granites. Total REE content in the granite, in general, is high with relative enrichment of LREE over HREE.

Early intrusive syntectonic granitic activity, involving anatexis and related migmatisation of pre-existing rocks forming gneissic granites and migmatites, took place in the Neoarchaean, for which Rb-Sr, whole-rock, isochron ages range from 2528 to 2659 Ma. Subsequently, the area witnessed the main granite tectonism in the Palaeoproterozoic, which culminated in the emplacement of pegmatites (host for rare-earth mineralization). The available whole-rock, Rb-Sr ages of granitic rocks from different parts of Bastar craton suggest remobilization of sialic crust that resulted in the emplacements of granitic rocks from an anatectic melt during the period ca. 2300-2100 Ma.

## ACKNOWLEDGMENTS

I express my sincere gratitude to Shri P.B. Maithani, Director, AMD, Hyderabad, for granting permission to contribute this chapter and to Ms. Sandhya Nagarale for secretarial assistance.

## REFERENCES

Anonymous (1978) New age data by $K^{(40)} - Ar^{(40)}$ method. Geol. Surv. *India News*, v. 9, 3p.

Babu, T.M. (1985) Geology, mineralogy and genetic characteristics of tin, niobium and tantalum mineralization of Katekalyan and adjoining areas of Bastar district, Madhya Pradesh. *Indian Minerals*, v. 39, pp. 1-8.

Babu, T.M. (1994) Tin in India. Geological Society of India, *Mineral Resource Series* 7, 217p.

Bandyopadhyay, B.K., Bhoskar, K.G., Ramachandra, H.M., Roy, A., Khadse, V. K., Mohan, M., Rao, K.S., Barman, T.R., Bishui, P.K. and Gupta, S.N. (1990) Recent geochronological studies in parts of the Precambrian of Central India. *Geol. Surv. India*, Spl. Pub., v. 28, pp. 199-210.

Bhattacharyya, B.P. (1988) Sequence of deformation, metamorphism and igneous intrusions in Bihar mica belt. *Geol. Soc. India*, Memoir 8, pp. 113-126.

Cerny, P. (1989) Characteristics of pegmatite deposits of tantalum. In: Moller, P., Cerny, P. and Saupe, F. (Editors), Lanthanides, Tantalum and Niobium. Springer Verlag, Spl. Pub.No7 of the Society for Geology Applied to Mineral Deposits, pp.195-239.

Cerny, P. (1992) Geochemical and petrogenetic features of mineralisation in rare-element granitic pegmatites in the light of current research. *Applied Geochemistry*, v. 7, pp. 393-416.

Crookshank, H. (1963) Geology of southern Bastar and Jeypore from Bailadila to Eastern Ghats. *Mem. Geol. Surv. India*, v. 87, 149p.

Dey, A.K. (1983). Geology and Mineral Resources of Jashpur, Raigarh district, Madhya Pradesh. *Memoir Geological Survey of India*, v. 114, 88p.

Dikshitulu, G.R. and Dhana Raju, R. (1997) Petrology and geochemistry of the granitoids of central gneissic complex in the Kameng district, Arunachal Pradesh. *Jour. Geol. Soc. India*, v. 50, pp. 407-419.

Gaines, R.V., Skinner, H.C.W., Foord, E.E., Mason, B. and Rosenzweig, A. (1997) Dana's New Mineralogy, 8th Edition, John Wiley and Sons, Inc. New York, 1819p.

Garrels, R.M. and Mackenzie, F.T. (1971). *Evolution of Sedimentary Rocks*. W.W. Norton, New York, 394 p.

Ghose, N.C. (1983) Geology, tectonics and evolution of the Chhotanagpur granite gneiss complex, eastern India. Recent Researches in Geology, v. 10, Hindustan Publ. Corp., New Delhi, pp. 211-247.

Ghose, N.C. (1992) Chhotanagpur gneiss-granulite complex, eastern India. Present status and future prospect. *Indian Jour. Geology*, v. 64, pp. 100-121.

Gupta, C.K. (1993) Extractive metallurgy of rare metals. *BARC News Letter* No. 112, pp. 1-16.

Gupta, C.K. and Krishnamurthy, N. (1990) Rare Earths-Their preparation and applications. *Indian Journal of Technology*, v. 28, pp. 247-258.

Harpum, J.R. (1963). Petrographic classification of granitic rocks by partial chemical analysis. *Tanganyika Geological Survey Report*, v. 10, pp. 80-86.

Hedrick, J.B. (1985) Rare-earth elements and yttrium. Mineral Facts and Problems. *Bureau of Mines Bulletin* 675, pp. 647-663.

Karkare, S.G. and Shrivastava, R.K. (1985) Geochemistry of Sukma granitoids. *Rec. Geol. Surv. India,* v. 116 (2), pp. 51-68.

Kuster, D. (1990) Rare-metal pegmatites of Wamba, central Nigeria - their formation in relationship to late Pan-African granites. *Mineral. Deposita*, v.25, pp. 25-33.

Lamba, V.S.S. and Agarkar, P.S. (1988) The tin potential of Precambrain rare-metal bearing pegmatites of Bastar, Madhya Pradesh, India. *Mineral. Deposita*, v. 23, pp. 218-221.

Lottermoser, B.G. (1991) Rare-earth element resources and exploration in Australia. *The Aus. IMM Proceedings*, No. 2, pp. 49-56.

Mahadevan, T.M. (1992) Geological evolution of the Chhotanagpur gneiss complex in a part of Purulia district, West Bengal. *Indian Jour. Geology*, v. 64, pp. 1-22.

Mahadevan, T.M. (2002) Geology of Bihar and Jharkhand. *Geol. Soc. India*, Bangalore, 563p.

Mahadevan, T.M. (2008) Precambrian geological and structural features of the Indian Peninsula. *Jour. Geol. Soc. India*, v. 72, pp. 35-55.

Mariano, A.N. (1989) Economic geology of rare earth elements. In: Lipin, B.R. and McKay, G.A. (Editors), Geochemistry and Mineralogy of Rare Earth Elements. *Mineralogical Society of America, Reviews in Mineralogy*, v. 21, pp. 309-337.

Maji, A.K., Goon, S., Bhattacharya, A., Mishra, B., Mahato, S. and Bernhardt, H. J. (2007) Proterozoic polyphase metamorphism in the Chhotanagpur Gneissic Complex (India), and implication for trans-continental Gondwanaland correlation. Precambrian Res., doi: 10.1016/j. precamres. 2007, 10.002.

Mazumdar, S.K. (1988) Crustal evolution of the Chhotanagpur gneissic complex and the mica belt of Bihar. *Geol. Soc. India*, Memoir 8, pp. 49-84.

Mishra, V.P., Singh, P. and Dutta, N.K. (1988) Stratigraphy, structure and metamorphic history of Bastar Craton. *Rec. Geol. Surv. India*, v. 117 (pt. 3-8), pp. 1-26.

Pandey, B.K., Prasad, R.N., Krishna, B., Sagar, S., Gupta, J.N. and Saraswat, A.C. (1989) Early Proterozoic Rb-Sr isochron ages for the granitic rocks from parts of Koraput district, Orissa, India. *Ind. Minerals*, v. 43, nos. 3-4, pp. 273-278.

Patel, S.C., Sundararaman, S., Dey, R., Thakur, S.S. and Kumar, M. (2007) Deformation pattern in a Proterozoic low pressure metamorphic belt near Ramanujganj, western Chhotanagpur terrane. *Jour. Geol. Soc. India*, v. 70, pp. 207-216.

Preinfalk, C. and Morteani, G. (1989) The industrial applications of rare earth elements. In: Moller, P., Cerny, P. and Saupe, F. (eds.), Lanthanides, Tantalum and Niobium. Springer-Verlag, pp. 359-370.

Rai, S.D. and Banerjee, D.C. (1995) Xenotime placers in parts of Mahan river basin, Surguja district, Madhya Pradesh. *Jour. Geol. Soc. India*, v. 45, pp. 285-293.

Rajasekharan, P., Senthilvel, C.N.S., Satyanarayana, K., Viswanathan, R., Reddy, L.S.R., Manjunath, Y.S. and Sinha, R.P. (1994) Bi-pyramidal xenotime-(Y) from Siri river placers, Raigarh district, Madhya Pradesh, India. *Jour. Atomic Mineral Science*, v. 2, pp. 1-17.

Ramakrishnan, M. and Vaidyanadhan, R. (2008) Geology of India. *Geol. Soc. India*, v.1, 556p.

Ramachar, T.M., Shivananda, S.R., Dwivedy, K.K. and Jayaram, K.M.V. (1983) The rare metal and REE occurrences in south Bastar, Madhya Pradesh. *Geol. Surv. India.* Spl. Publ. No. 13, pp. 104-107.

Ramesh Babu, P.V. (1999) Rare metal and rare earth pegmatites of Central India. *Exploration and Research for Atomic Minerals*, v. 12, pp. 7-52.

Ramesh Babu, P.V., Dwivedy, K.K. and Jayaram, K.M.V. (1984) Geochemistry of tin-rich granites of Paliam and Darba, Bastar district, M.P. Proc. 5th Indian Geological Congress, Bombay, pp. 313-319.

Ramesh Babu. P.V., Pandey, B.K. and Dhana Raju, R. (1993) Rb-Sr ages on the granite and pegmatitic minerals from Bastar-Koraput pegmatite belt, Madhya Pradesh and Orissa, India. *Jour. Geol. Soc. India*, v. 42, pp. 33-38.

Ramesh Babu, P.V., Rajendran, R., Vijaya Kumar, K., Singh, R.P. and Banerjee, D.C. (1998) Exploration for xenotime placers and their source rocks in Balrampur-Bhandaria area, Surguja district, Madhya Pradesh and Garhwa district, Bihar, India. *Exploration and Research for Atomic Minerals*, v. 11, pp. 37-44.

Sarah, L.H., Simmons, W.B. and Webber, K.L. (1992) Rare earth element mineralogy of granitic pegmatites in the Trout Creek Pass district, Chaffeo County, Colorado. *Canadian Mineralogist*, v. 30, pp. 673-686.

Sarkar, G., Paul, D.K., DeLaeter, J.R., McNaughton, N.J. and Mishra, V.P. (1990a). A geochemical and Pb, Sr isotopic study of the evolution of granite-gneisses from the Bastar craton, central India. *Jour. Geol. Soc. India*, v. 35, no. 5, pp. 480-496.

Sarkar, A., Sarkar, G., Paul, D.K. and Mitra, N.D. (1990b) Precambrian geochronology of the Central Indian Shield – a review. *Geol. Surv. India*, Spl. Publ. No. 28, pp. 453-482.

Sarkar, A.N. (1982) Precambrian tectonic evolution of eastern India: a model of converging microplates. *Tectonophysics*, v. 86, pp. 363-397.

Sarkar, A.N. (1988) Tectonic evolution of the Chhotanagpur plateau and the Gondwana basins in eastern India: an interpretation based on supra-subduction geological processes. *Geol. Soc. India*, Memoir 8, pp. 127-146.

Saxena, V.P., Krishnamurthy, P., Murugan, C. and Sabot, H.K. (1992) Geochemistry of the granitoids from the central Surguja shear zone, India: geological evolution and implications on uranium mineralisation and exploration. *Exploration and Research for Atomic Minerals*, v. 5, pp. 27-40.

Singh, Yamuna (1990) Identification of xenotime-bearing Precambrian granites in central India and its economic significance. Proc. National Seminar held at Geology Department, Andhra University, Visakhapatnam, during March 25-26, 1990, Abstract Volume, pp. 36-37.

Singh, Yamuna (1991) Precambrian rare metal pegmatites near Belangi, central India. Proc. 2nd Conference on Geology of Indo-China, 11-12 November, 1991, Hanoi, v. 1, pp. 281-290.

Singh, Yamuna (1992). Geochemistry of porphyritic granite from Kunkuri-Rajauti area, Raigarh district, Madhya Pradesh. *Indian Jour. Geology*, v. 64, pp. 23-36.

Singh, Yamuna (1998) Early Proterozoic rare metal and tin pegmatites near Kawadgaon, Bastar, M.P: An example of vertical pegmatite zonation. *Jour. Geol. Soc. India*, v. 51, pp. 175-182.

Singh, Yamuna (1999) Lithostratigraphic correlation of andalusite quartzites above the early Proterozoic granites intrusive into Bengpal and Sukma Groups in Gadapal-Jaram area, South Bastar, Madhya Pradesh. *Gond. Geol. Mag.*, v. 14, No. 2, pp. 1-9.

Singh, Yamuna (2002) Fluvial landform, mineralogical and geochemical (Y, Ce, Nb, Ta, Sn) studies on grantic soil/alluvial deposits in the environs of Kawara and Nakti sub-basins around Amraoti, Bastar district, Chhattisgarh. *Jour. Indian Assoc. Sedimentologists*, v. 21, No. 1and2, pp. 1-13.

Singh, Yamuna (2007) Granitic soil yttrium-cerium geochemistry with xenotime concentrate upgradation around Darba, Bastar district, Chhattisgarh. *Gond. Geol. Mag.*, Special Volume No. 10, pp. 173-178.

Singh, Yamuna and Rai, S.D. (1992) Yttrium-europium-erbium geochemistry of granitic soils of Kunkuri area, Raigarh district, M.P., India. *Jour. Geol. Soc. India*, v. 40, pp. 347-358.

Singh, Yamuna and Singh, K.D.P. (1996) Mineralogical and geochemical studies on xenotime-bearing granites of Raikera area, Raigarh district, Madhya Pradesh, India. *Jour. Atomic Mineral Science*, v. 4, pp. 37-47.

Singh, Yamuna and Chabria, T. (1999) Late Archaean-early Proterozoic Rb-Sr isochron age of granite from Kawadgaon, Bastar district, Madhya Pradesh. *Jour. Geol. Soc. India*, v. 54, no. 4, pp. 405-409.

Singh, Yamuna and Singh, K.D.P. (2001) X-ray mineralogy and cerium-yttrium-niobium-tantalum geochemistry of granitic soil from Muskel-Kawadgaon area, Bastar Craton, Central India. *Indian Minerals*, v. 55, No. 1 and 2, pp. 47-54.

Singh, Yamuna and Chabria, T. (2002) Early Proterozoic $^{87}$Rb-$^{86}$Sr model ages of pegmatite muscovite from rare metal-bearing granite-pegmatite system of Kawadgaon, Bastar craton, central India. *Gondwana Res.*, v. 5, no. 4, pp. 889-893.

Singh, Yamuna and Rai, S.D. (2008) Geochemistry of pink granite from the Precambrian Chhotanagpur granite gneiss complex in a part of Gumla district, Jharkhand: an example of alkali metasomatism. *Geol. Soc. India,* Memoir 73, pp. 85-100.

Singh, Yamuna and Reddy, L.S.R. (2008) Petrology of the REE- and Y-bearing granitoids from the Raikera-Kunkuri area, Central India. *Exploration and Research for Atomic Minerals*, v. 18, pp. 55-76.

Singh, Yamuna and Krishna, V. (2009) Rb-Sr geochronology and petrogenesis of granitoids from the Chhotanagpur granite gneiss complex of Raikera-Kunkuri region, Central India. *Jour. Geol. Soc. India*, v. 74, pp. 200-208.

Singh, Yamuna and Shiv Kumar, K. (2009) A note on radioactive riverine placers around Midtul, Bastar Craton, Central India. *Jour. Geol. Soc. India*, v. 73, pp. 419-424.

Singh, Yamuna, Parihar, P.S. and Maithani, P.B. (2011) Chemistry of thorium-rich monazite-(Ce) associated with the rare-metal-mineralised Precambrian pegmatites in Central India. Proc. Intl. Symposium on Precambrian Accretionary Orogens. Abstract Volume, pp. 157-159 (Published by Geological Society of India, Bangalore).

Spooner, J., Grace, K.A. and Robjohns, N. (1991) The economics of the rare-earth elements. *CIM Bulletin*, v. 84 (947), pp. 125-131.

Streckeisen, A. (1976) To each plutonic rock its proper name. *Earth Sci. Rev.*, v. 12, pp. 1-33.

Thuy Nguyen, T.B., Satir, M., Siebel, W., Vennemann, T. and Long, T.V. (2004) Geochemical and isotopic constraints on the petrogenesis of granitoids from the Dalat zone, Southern Vietnam. *Jour. Asian Earth Sciences*, v. 23, pp. 467-482.

Winter, J.D. (2001) An Introduction to Igneous and Metamorphic Petrology. Prentice Hall, New Jersey, 697p.

*Chapter 2*

# OCCURRENCE OF TH, U, Y, ZR, AND REE-BEARING ACCESSORY MINERALS IN GRANITES AND THEIR PETROGENETIC SIGNIFICANCE

### *Miloš René*
Institute of Rock Structure and Mechanics,
Academy of Sciences of the Czech Republic,
Czech Republic

### ABSTRACT

The association of Th, U, Zr, Y, and REE accessory minerals (zircon, monazite, xenotime) in two-mica granites and topaz granites of the Saxo-Danubian granite belt (Central European Hercynides) are discussed. The occurrence of these minerals is controlled by the initial trace-element contents of the melt, the aluminous saturation index, the Ca content, as well as F, P and Y contents, and the LREE/HREE and U/Th ratios of the melt. In topaz granites the late-magmatic and post-magmatic hydrothermal processes these assemblages and their compositions also control. The ca. 400 km long plutonic megastructure of the Saxo-Danubian belt is formed by Fichtelgebirge/Erzgebirge compositional batholith in the Saxothuringian Zone and the South Bohemian batholith (SBB) in the Moldanubian Zone of the Central European Hercynian fold belt. Two-mica granites form an important part of the SBB. Two, mineralogically and geochemically contrasting granites, the Eisgarn and the Deštná granites, represent the majority of these granites. In the Krušné Hory/Erzgebirge area according to their geochemical signatures P-rich and P-poor topaz granites can be distinguished.

In the Eisgarn granite accessory minerals are concentrated in biotite flakes, whereas in the Deštná granite accessory minerals are predominantly enclosed in K-feldspar phenocrysts. These differences together with different Zr, Th and LREE concentrations in both granite types indicate quite different melting histories of both granites melts. Thorium, U, Y, Zr and REE-bearing accessory minerals in both two-mica granite types are represented by apatite, monazite, and zircon. In addition, in the Deštná granite rather rare xenotime-(Y) occurs. Monazite in both granite types is Ce-dominant (1.3–1.8 apfu Ce) with a restricted amount of cheralite component. However, monazite from the Deštná granite is partly enriched in Y (up to 8.6 mol.% $YPO_4$). Xenotime-(Y) from the Deštná granite has a composition along the solid solution between xenotime and cheralite with predominance of U over Th (Th/U = 0.08–0.33).

For topaz granites zircon–xenotime intergrowths are significant. Zircon from the P-rich topaz granites displays a significant enrichment in P (up to 8.3 wt.% $P_2O_5$) and depletion in Y compared to zircon from the P-poor granites. Zircon from the P-poor topaz granites contains up to 18.4 wt.% $Y_2O_3$. Xenotime from the P-poor granites also displays a considerable enrichment in HREE (up to 35.7 wt.% $HREE_2O_3$) compared to xenotime from P-rich granites (up to 19.5 wt.% $HREE_2O_3$). For xenotime from P-rich granites the brabantite-type substitution is particularly significant, whereas in xenotime from P-poor granites the thorite-type substitution occurred.

# INTRODUCTION

In the last twenty years, several studies have emphasized the role of accessory minerals as major hosts for REE, Th, U, Zr and Y elements in granites [1, 2, 3, 4, 5, 6, 7]. The crystallization of accessory minerals, including monazite, xenotime, zircon and apatite is controlled by parameters that include $SiO_2$ activity, aluminous saturation index (ASI), the ratios and concentrations of REE, Th, U, Zr, and Y, as well as the oxygen fugacity and Ca, Fe, Mg content of the melt. This paper presents compositional and petrological observations on the occurrence of zircon, monazite and xenotime assemblage in two contrast two-mica granite types of the South Bohemian Batholith and zircon-xenotime assemblage in topaz granites from the Krušné Hory/Erzgebirge area. Both granite groups are part of Saxo-Danubian granite belt, which represent one from five bigger granite belts of the Central European Hercynian granite belt.

# GEOLOGICAL BACKGROUND

The European Hercynian orogen is a collage of microplates that were assembled, between the Devonian and the Carboniferous, along the southern margin of the Old Red Continent [8, 9, 10]. The central European section of the Hercynian orogen, with the Bohemian Massif as the main exposure, includes several independent collision zones that represent fold belts active at different times. During the Carboniferous, numerous granitic plutons intruded all over the Bohemian Massif. These are commonly treated as a single coherent group of "Hercynian (Variscan) granites" [8], although they are of different ages and types. It is suggested that they form at least five independent magmatic systems with individual tectonothermal backgrounds (Figure 1).

The southwestern sector of the Bohemian Massif was intruded by numerous, mainly crustally derived, granitic magmas between ca 330 and 310 Ma, associated with penetrative LP-HT regional metamorphism and anatexis. The South Bohemian Batholith forms granitic complex of batholithic dimensions developed at the eastern end of the Bavarian zone. The SBB form a coherent plutonic belt with various granite plutons evolved in Smrčiny/Fichtelgebirge and Krušné Hory/Erzgebirge areas. For this big granite belt, Finger et al. [11] proposed a new denomination, the Saxo-Danubian Granite Belt. Several studies in the Krušné Hory/Erzgebirge area [12], the Smrčiny/Fichtelgebirge Mts. and Oberpfalz Forest

[13], the Bavarian Forest [14] and the South Bohemian Batholith [15] have pointed out that the majority of the granites from these areas are derived through fluid-absent partial melting of lower crustal sources.

In the Krušné Hory/Erzgebirge area, topaz granites associated with Sn-W ore deposits play an important role. Their presence may reflect a local availability of Sn-enriched crustal source rocks [12]. Other differences between the Krušné Hory/Erzgebirge granites and the South Bohemian Batholith appear to be a matter of a different exposure level. Many granitic intrusions in the Krušné Hory/Erzgebirge area are felsic high-level plutons with a high degree of fractionation, while the exposure level of the South Bohemian Batholith is generally deeper and involves considerable amounts of cumulate material [16].

The South Bohemian Batholith granite complex of the central and southern part of the Moldanubian Zone, exposes n over 6,000 km² of plutonic rocks, and is one of the largest plutonic bodies in the European Variscan fold belt. The SBB consists of multiple plutons and comprises a number of petrographic types. The individual granite bodies of the batholith are clustered in two nearly perpendicular oriented ~ NNE–SSW and ~ WNW–ESE segments. While the ~ WNW–ESE (Šumava Mts./Böhmerwald) segment is defined by a number of separate smaller granite bodies, the ~ NNE–SSW segment is chiefly made up of one bigger pluton (Central Moldanubian Pluton, CMP) (Figure 2).

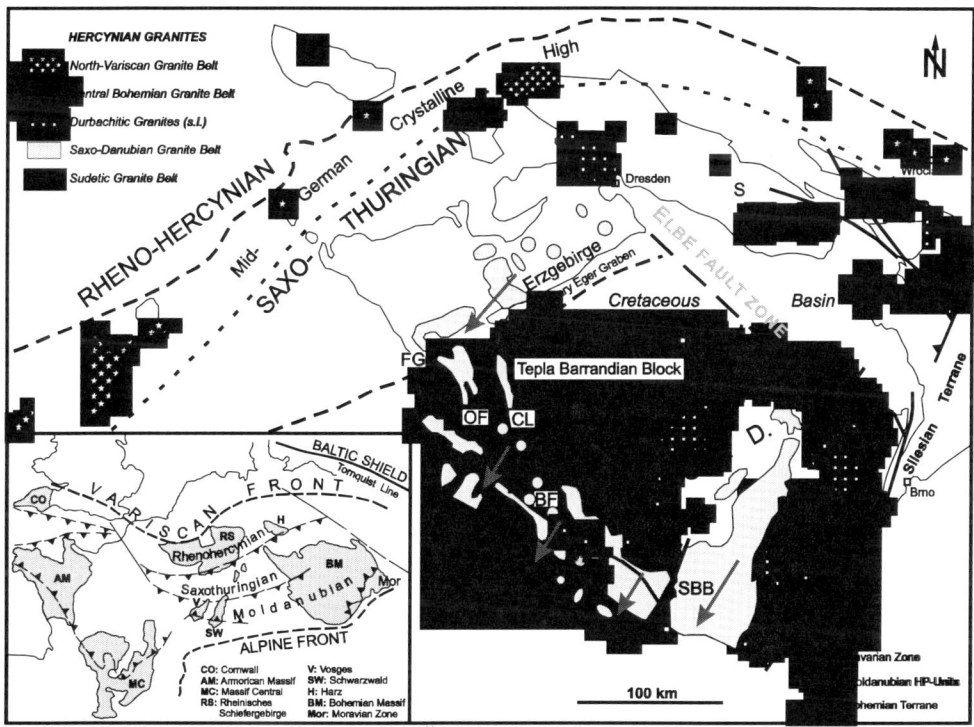

Abbreviations: BF=Bavarian Forest, CL=Český Les, FG=Fichtelgebirge, OF=Oberpfalz Forest, RF=Regensburg Forest, SBB=South Bohemian Batholith, S=Sudetes, SW=Sauwald. Inset illustrates the position of the Bohemian Massif within the entral European Hercynides.

Figure 1. Sketch map of the Bohemian Massif, mainly after Franke [8], showing the distribution of Hercynian granites. Arrows indicate the proposed growth direction of the SDGB.

Granitic rocks of the SBB can be divided into three main intrusive suites: (1) coarse-grained K-feldspar-porphyric biotite granite (the Weinsberg type), (2) peraluminous two-mica granite and (3) equigranular, fine grained biotite granite to granodiorite (the Mauthausen and Freistadt types) [15, 18]. Two-mica granites of the SBB are represented by four, texturally, mineralogically and geochemically different types [19, 20, 21], namely the Eisgarn, Deštná, Lipnice (Steinberg) and Zvůle (Čeřínek, Melechov) types. However, the Eisgarn and the Deštná types represent the majority of the two mica granites in the SBB.

The Lipnice granite occurs only in the Melechov granite body, while the Steinberg granite occurs in the Třístoličník/Dreisessel granite stock [17]. The Zvůle, Čeřínek and Melechov types form three independent granite stocks within the CMP (Figure 2).

Topaz granites form two distinct magmatic suites in the Krušné Hory/Erzgebirge area [12, 22, 23, 24] (Figure 3). The first suite is a highly evolved P-rich (>0.4 wt.% $P_2O_5$) S-type granite.

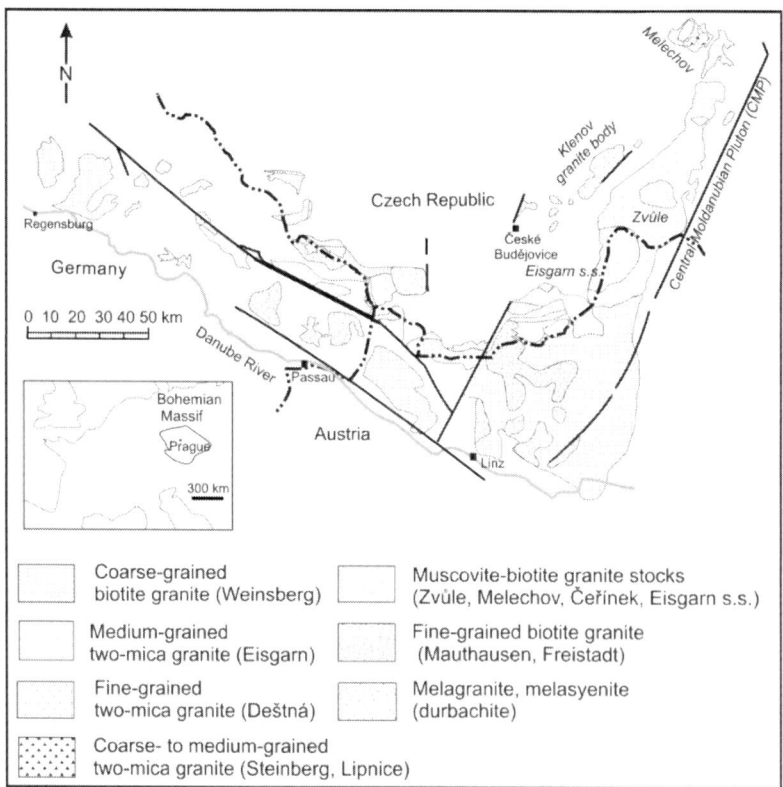

Figure 2. Geological map of the South Bohemian (Moldanubian) batholith (after Siebel et al. [14] and Verner et al. [17], modified by author).

High-$P_2O_5$ topaz-bearing granites in sense of Taylor and Fallick [26] occur in granite stocks of the western and central parts of the Krušné Hory/Erzgebirge. The most important occurrences of the P-rich granite suite are stocks in the area of the Ehrenfriedersdorf Sn–W ore deposit (Germany) and in the Horní Slavkov–Krásno ore district (Czech Republic). The second topaz-bearing granite suite is formed by P-poor (<0.4 wt.% $P_2O_5$), A-type granites. These granites include small granite bodies and stocks with important Sn–W ore mineralization of the volcanic-plutonic complex of the eastern Krušné Hory/Erzgebirge area (Altenberg, Zinnwald/Cínovec, Sadisdorf, Sachsenhöhe, Krupka).

The Horní Slavkov–Krásno ore district comprises mineralized topaz-bearing granite stocks along the SE margin of the Krudum granite body in the Slavkovský les Mts. area. [27]. The inner structure of granite stocks (Hub, Schnöd, Vysoký Kámen) are well stratified, comprising partly greisenized topaz-albite granites, leucocratic topaz-albite granites and layers of alkali-feldspar syenites. In the upper part of the Hub stock, a pipe of gneissic breccia cemented by topaz-albite microgranites also occurs. Leucocratic topaz-albite granite together with alkali-feldspar syenite forms the Vysoký Kámen stock.

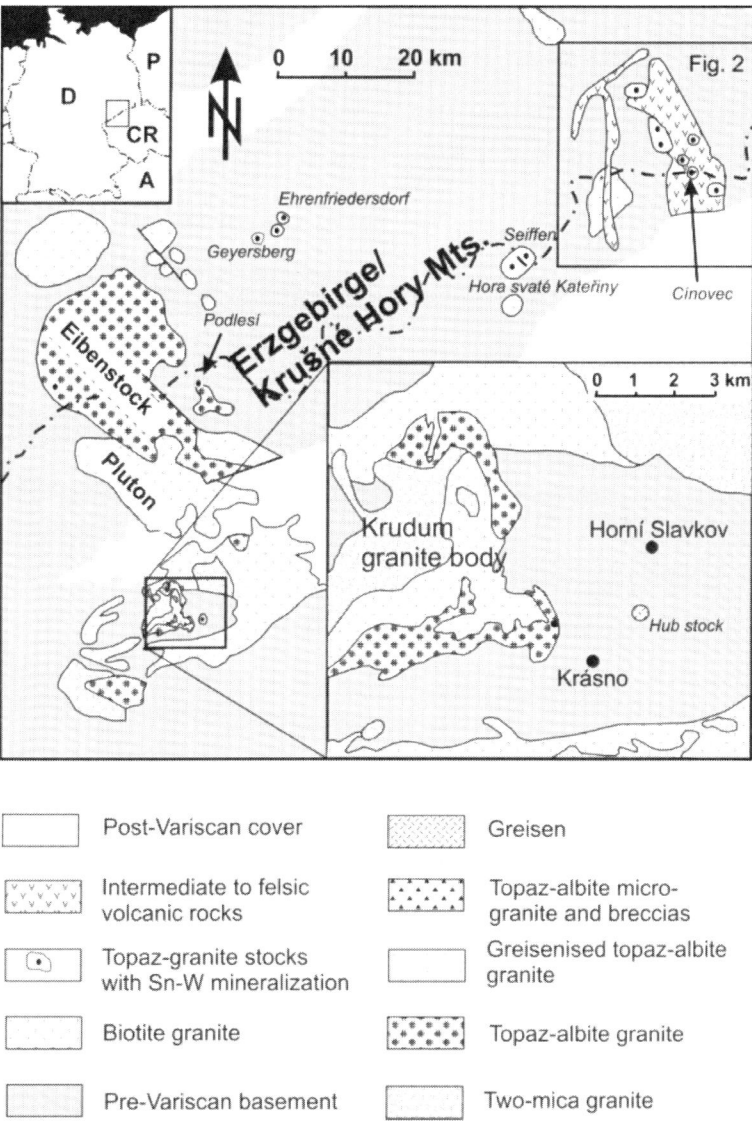

Figure 3. Distribution of late-Hercynian igneous rocks in the Krušné Hory/Erzgebirge area, with the Krudum granite body and associated granite stocks shown as inset map (modified from Müller et al. 2003 [25]).

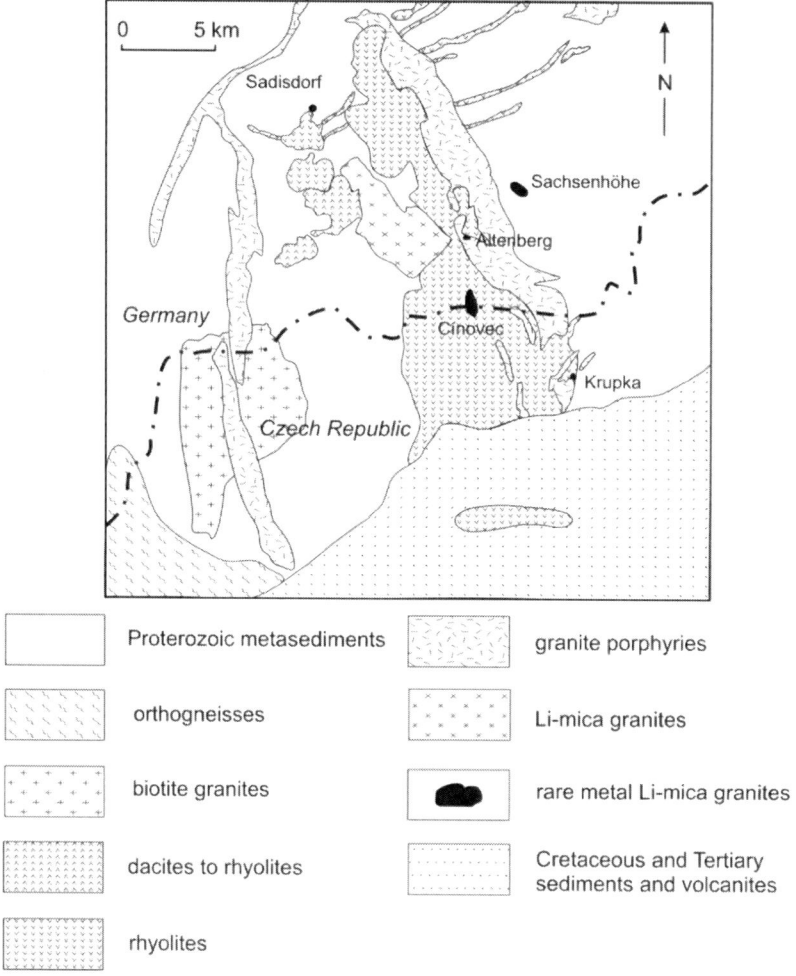

Figure 4. Geological map of the volcano-plutonic complex of the eastern Krušné Hory/Erzgebirge with location of the Cínovec granite stock (after Seifert and Kempe [28], modified by author).

The Cínovec granite stock is a relatively small, elliptical, vertical stratified body in the central part of the Altenberg-Teplice caldera (Figure 4).

The borehole CS-1, located in the center of granite stock, transacted lepidolite-bearing granite at the top of the section (about 90 m thick), an intermediate zone of zinnwaldite-bearing granite (thickness about 640 m) and a lower zone of protolithionite-bearing granite to the depth 1596 m [29, 30].

## ANALYTICAL METHODS

Approximately 300 quantitative electronprobe microanalyses of zircon, monazite and xenotime in representative samples of two-mica granites from SBB and 200 quantitative electronprobe analyses of zircon and xenotime in representative topaz granite samples from the Krušné Hory/Erzgebirge were performed. Minerals were analyzed in polished thin sections and back-scattered electron images (BSE) were acquired to study interaction of examined accessory minerals and the internal structure of individual mineral grains. Element abundances of Al, As, Bi, Ca, Dy, Er, F, Fe, Gd, Hf, Ho, La, Lu, Mg, Mn, Nb, Nd, P, Pb, Pr, Sc, Si, Sm, Ta, Th, U, W, Y, Yb and Zr were determined using a CAMECA SX 100 electron microprobe operated in wavelength-dispersive mode at the Institute of Geological Sciences, Masaryk University in Brno. The accelerating voltage and beam currents were 15 kV and 20 or 40 nA, respectively, with beam diameter 1 to 5 μm. The following standards, X-ray lines, and crystals (in parentheses) were used: $K_\alpha$: Al on almandine (TAP), Ca and Fe on andradite (LLIF), F on topaz (PC1), Mn on rhodonite (LLIF), Mg on olivine (TAP), P on fluorapatite (PET), Sc on $ScVO_4$ (PET), Si on zircon (TAP), $L_\alpha$: As on InAs (TAP), Ce on $CePO_4$ (PET), Gd – $GdF_3$ (LLIF), La on $LaPO_4$ (PET), Nb on ferrocolumbite (LPET), Sm on $SmPO_4$ (LLIF), W on scheelite (LLIF), Y on YAG (TAP), Yb on $YbP_5O_{14}$ (LLIF), Zr on zircon (TAP), $L_\beta$: Dy on $DyPO_4$ (LLIF), Er on YErAG (LLIF), Ho on $HoPO_4$ (LLIF), Nd on $NdF_3$ (LLIF), Pr on $PrF_3$ (LLIF), $M_\alpha$: Bi on metallic Bi (LPET), Hf on $HfO_2$ (TAP), Pb on PbSe (LPET), Th on $ThO_2$ (LPET); and $M_\beta$: Lu on $Lu_3Al_5O_{12}$ (TAP), U on metallic U (LPET). Peak count-time was 20 s and background time 10 s for major elements, whereas for trace elements they were 40–60 s and 20–30 s, respectively. The raw data were corrected using PAP matrix corrections [31]. Detection limits were approximately 400–500 ppm for Y, 600 ppm for Zr, 500–800 ppm for REE and 600–700 ppm for U and Th. Formulae of monazite and xenotime were calculated in relation to 8 oxygen atoms per formula unit (apfu) and formulae of zircon were calculated in relation to 4 oxygen atoms per apfu.

The whole rock composition of the Eisgarn and Deštná two-mica granites is based on 52 samples. The whole rock composition of topaz granites was determined for 46 representative samples from the Horní Slavkov–Krásno ore district, the Podlesí, Geyersberg, Sauberg and Cínovec granite stocks. Major elements were determined by X-ray fluorescence spectrometry using the PANanalytical Axios Advanced spectrometer at Activation Laboratories Ltd., Ancaster, Canada. The content of FeO was measured by titration, whereas the content of water was determined gravimetrically at the analytical laboratory of the Institute of Rock Structure and Mechanics AS CR, v.v.i., Prague, Czech Republic. The content of F was determined by using an ion-selective electrode in the same laboratory. Trace elements were determined by ICP MS (a Perkin Elmer Sciex ELAN 6100 ICP mass spectrometer) at Activation Laboratories Ltd., Ancaster, Canada. Since the analytical procedure for ICP MS involves using a lithium metaborate/tetraborate flux fusion, the Li concentration was analyzed separately by atomic absorption spectrometry on a Varian 220 spectrometer at the Institute of Rock Structure and Mechanics, AS CR, v.v.i, Prague, Czech Republic.

# REGIONAL PETROLOGY

## Two-Mica Granites of the SBB

The two-mica granites of the Eisgarn type are represented by two structural subtypes, which also display a variable ratio of muscovite and biotite. The main Eisgarn/Čímeř subtype is a porphyric, medium-grained, occasionally coarse-grained two-mica granite containing K-feldspar (14–50 vol.%), quartz (22–41 vol.%), plagioclase ($An_{5-25}$) (8–38 vol.%), biotite (3–16 vol.%) and muscovite (1–8 vol.%) assemblage. K-feldspar phenocrysts are between 10 and

30 mm in size and rarely exceed 5% by volume. The biotite usually predominated over muscovite. The Mrákotín/Sulzberg subtype is usually an equigranular, fine- to medium grained two-mica granite, with rare K-feldspar phenocrysts, containing K-feldspar (14–42 vol.%), quartz (23–46 vol.%), plagioclase (10–37 vol.%), muscovite (3–17 vol.%) and biotite (3–11 vol. %). Apatite, ilmenite, andalusite, zircon and monazite are common accessory minerals in both structural varieties.

The equigranular and or slightly porphyric, fine-grained Deštná type predominantly occurs in the Klenov granite body. This biotite-muscovite granite contains K-feldspar (21–47 vol.%), quartz (28–42 vol.%), plagioclase ($An_{7-24}$) (15–29 vol.%), muscovite (1–8 vol.%) and biotite (1–5 vol.%); the ratio of muscovite to biotite is typically >1. In this type the occurrence of schlieren or small nodular aggregates of restitic biotite is significant. Its accessory mineral assemblage consists of apatite, andalusite, ilmenite, monazite, zircon, and rare xenotime.

The above-mentioned types of two-mica granites could be also recognized on the basis of a different $Al_2O_3/TiO_2$ ratio (Figure 5 a) and their contents of Rb, Zr, Th and REE. Both granite types are peraluminous rocks with an aluminous saturation index (A/CNK) ranging from 1.1 to 1.3.

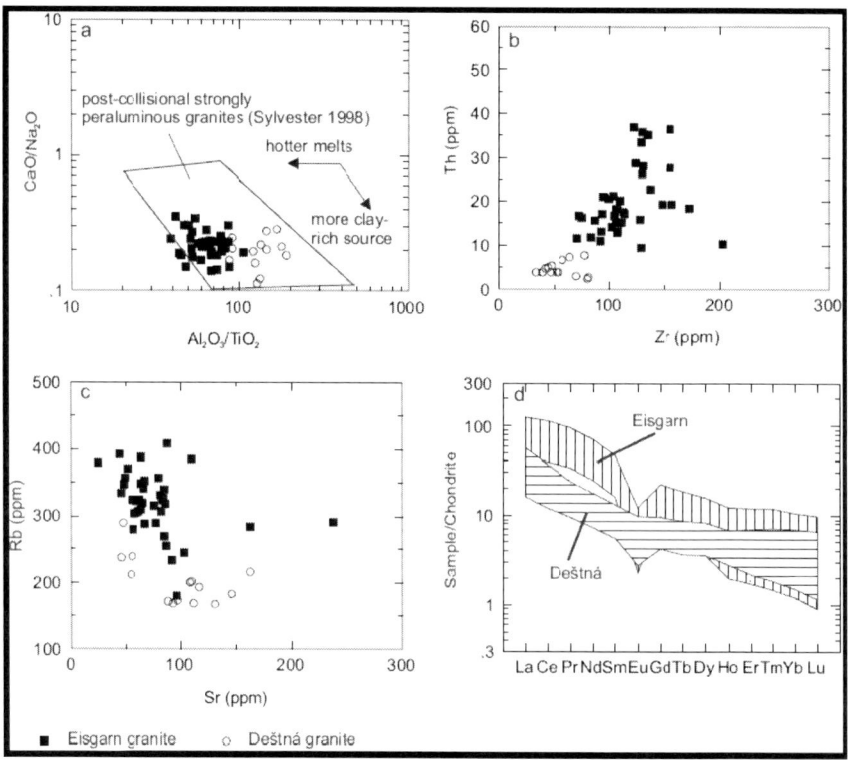

Figure 5. Geochemical features of two-mica granites of the South Bohemian Batholith, a) Plot according Sylvester [33], b) Plot of Th vs. Zr, c) Plot of Rb vs. Sr, d) Chondrite normalized REE patterns. Normalizing values are from Taylor and McLennan [34].

Compared to common Ca-poor granites [32], the Eisgarn granite is enriched in incompatible elements such as Rb (180–410 ppm), Cs (2–33 ppm), Sn (5–25 ppm), Nb (5–21 ppm), Zr (75–180 ppm) and Th (8–37 ppm) (Fig 5 b) and poor in Mg (0.2–0.7 wt.% MgO), Ca (0.4–1.2 wt.% CaO) and Sr (25–110 ppm) (Figure 5 c).

The Eisgarn granites are characterized by higher ratio LREE/HREE (La$_N$/Yb$_N$ = 20–48) and a particularly remarkable negative Eu anomaly (Eu/Eu* = 0.16–0.48). The Deštná type is depleted in Mg (0.2–0.4 wt.% MgO), Th (2–7 ppm) and Zr (15–80 ppm) and bulk of REE (33-69 ppm) relative to the Eisgarn granite. The Deštná granite has a characteristicly lower LREE/HREE (La$_N$/Yb$_N$ = 3–11) ratio and a rather positive Eu anomaly (Eu/Eu* = 0.81–1.17) (Figure 5 d).

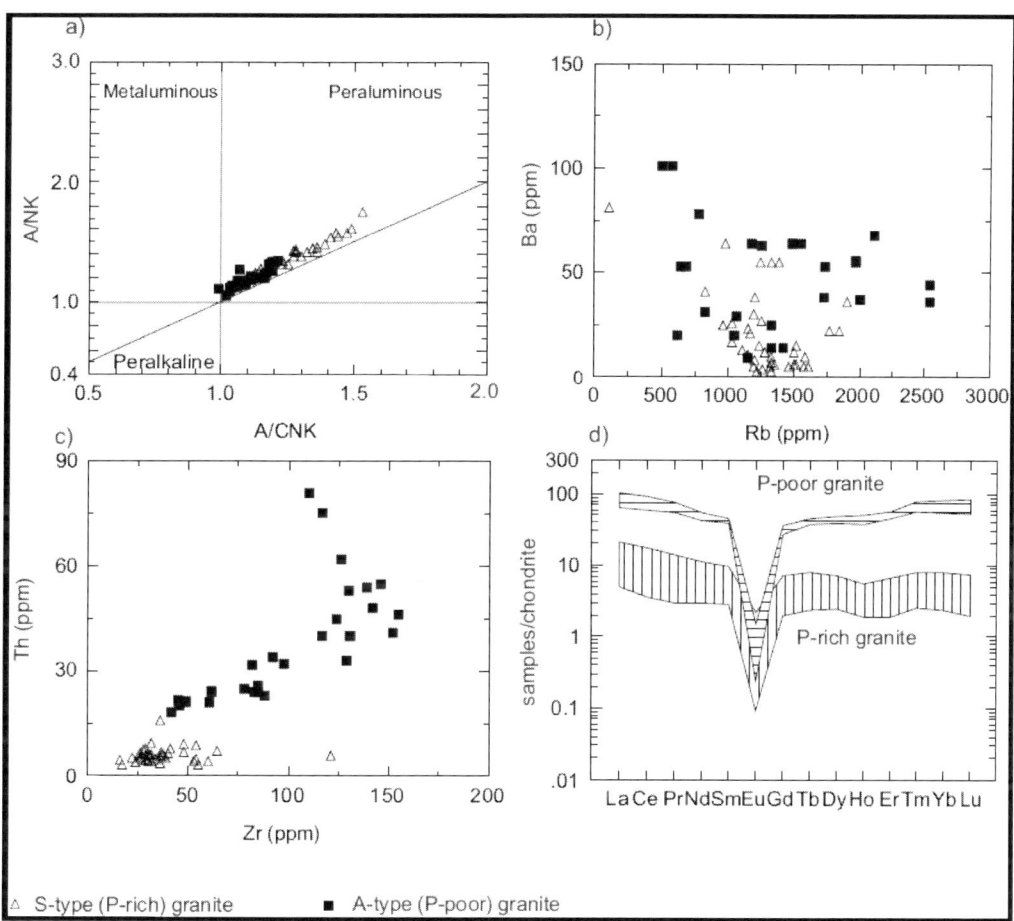

Figure 6. Geochemical features of topaz granites from the Krušné Hory/Erzgebirge area. a) Shand´s diagram after Maniar and Piccoli [35], b) Plot of Rb vs. Ba, c) Plot of Th vs. Zr, d) Chondrite normalized REE patterns. Normalizing values are from Taylor and McLennan [34].

## Horní Slavkov–Krásno Ore District

The partly greisenized topaz-albite alkali feldspar granite (TAG) is a medium-grained, equigranular rock, which consists of quartz, albite (An$_{0-2}$), potassium feldspar, lithium mica (protolithionite) and topaz. Fluorapatite, zircon, Nb-Ta-Ti oxides, xenotime-(Y), monazite-

(Ce), uraninite and coffinite are common accessory minerals. TAG is a peraluminous rock with an aluminous saturation index (A/CNK) ranging from 1.1 to 1.5 (Figure 6 a).

Compared to common Ca-poor granites [32], it is enriched in incompatible elements such as Li (160–820 ppm), Rb (830–1500 ppm), Cs (38–150 ppm), Sn (19–6200 ppm), Nb (18–83 ppm), Ta (8–53 ppm), W (4–62 ppm) and poor in Mg (0.1–0.2 wt.% MgO), Ca (0.3-1.0 wt.% CaO), Sr (12–50 ppm), Ba (21–81 ppm) and Zr (20–55 ppm) (Figure 6 b, c). The granite is distinctly enriched in phosphorus (0.3–0.4 wt.% $P_2O_5$) and fluorine (0.1–0.8 wt.% F). Granite is distinctly depleted in REE ($\Sigma REE$ = 12–46 ppm) (Figure 6 d). A high degree of magmatic fractionation is reflected in the low K/Rb ratio [15–47].

## Cínovec Granite Cupola

The zinnwaldite-bearing granite is predominantly a medium grained rock containing quartz, albite ($An_{0-5}$), potassium feldspar and zinnwaldite. The topaz, fluorite, Nb-Ta-Ti-oxides, cassiterite, zircon, Th-rich monazite-(Ce), bastnäsite, REE oxyfluorides and hydroxyfluorides, pyrochlore, wolframite and scheelite represent the accessory mineral assemblage. The granite is a slightly peraluminous rock with A/CNK ranging from 1.1 to 1.2 (Figure 6 a). Compared to typical A-type granites [36], it is enriched in Rb (584–2371 ppm), Nb (56–86 ppm), Ga (27–58 ppm) and depleted in Zr (24–98 ppm), Y (42–122 ppm) and Ce (81–87 ppm) (Figure 6 b, c).

Granite is distinctly enriched in REE ($\Sigma REE$ = 226–251 ppm) (Figure 6 d). A high degree of magmatic fractionation is reflected in the low K/Rb ratio [14–32]. Slightly porphyric protolithionite-bearing syenogranite contain quartz, albite ($An_{5-10}$), potassium feldspar and protolithionite. Topaz, zircon, xenotime-(Y), thorite, synchysite-(Y), synchysite-(Ce), Th-poor monazite-(Ce) and rutile are accessories. This granite is also slightly peraluminous with A/CNK ranging from 1.0 to 1.1 (Figure 6 a).

Compared with zinnwaldite-bearing granite, it is depleted in Rb (615–830 ppm), Nb (41–57 ppm) and Ga (25–26 ppm), but it is enriched in Zr (82–146 ppm), Y (116–149 ppm) and Th (32–75 ppm) (Figure 6 b, c). Granite is also distinctly enriched in REE ($\Sigma REE$ = 190–245 ppm) (Figure 6 d). Its lower degree of magmatic fractionation is reflected in a higher K/Rb ratio [52–66].

## ACCESSORY MINERALS ASSOCIATION

## Two-Mica Granites of the SBB

In the Eisgarn granite, zircon and monazite, together with fertile apatite and ilmenite are usually concentrated in biotite flakes (Figure 7 a, b, c, d). Anhedral apatite grains (100–600 µm) also contain abundant zircon and monazite inclusions (Figure 7 c). Euhedral to subhedral zircons and rarer anhedral monazite occur often on rims of larger apatite grains (Figure 7 a, b). The oscillatory zoning of zircon is rather scarce (Figure 7 e). Older inherited zircon cores overgrowth by younger magmatic zircon are also rare (Figure 7 f).

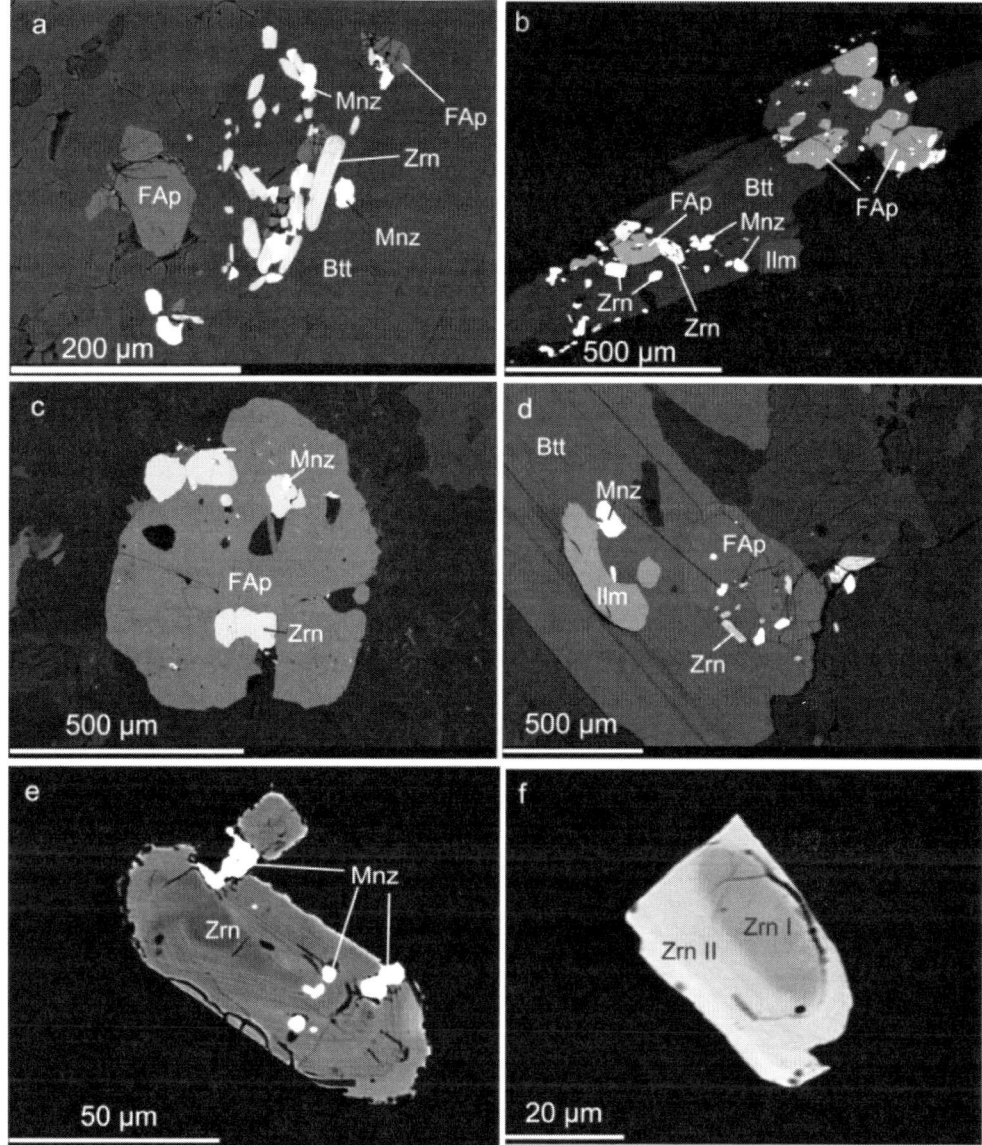

Figure 7. High contrast BSE images of accessory minerals from two-mica granites of the Eisgarn type (South Bohemian Batholith). Mineral abbreviations according to Whitney and Evans [56].

In the Deštná granite these accessories are distinctly rare and are preferentially enclosed in K-feldspar phenocrysts (Figure 8 a, b). All zircon and monazite grains in the Deštná granite are usually very small in size (5–15 µm).

Rare anhedral xenotime grains also occur in this granite (5–10 µm) (Figure 8 b, c, d). Some of these grains are enclosed in zircon and/or partly bigger xenotime grains contain very small zircon grains (Figure 8 c, d). Apatite in the Deštná granite forms are relatively small (20-100 µm), usually inclusion-free anhedral grains (Figure 8 b, d). Monazite and zircon grains in the Deštná granite also form irregular intergrowths (Figure 8 e) and/or are enclosed in ilmenite (Figure 8 f).

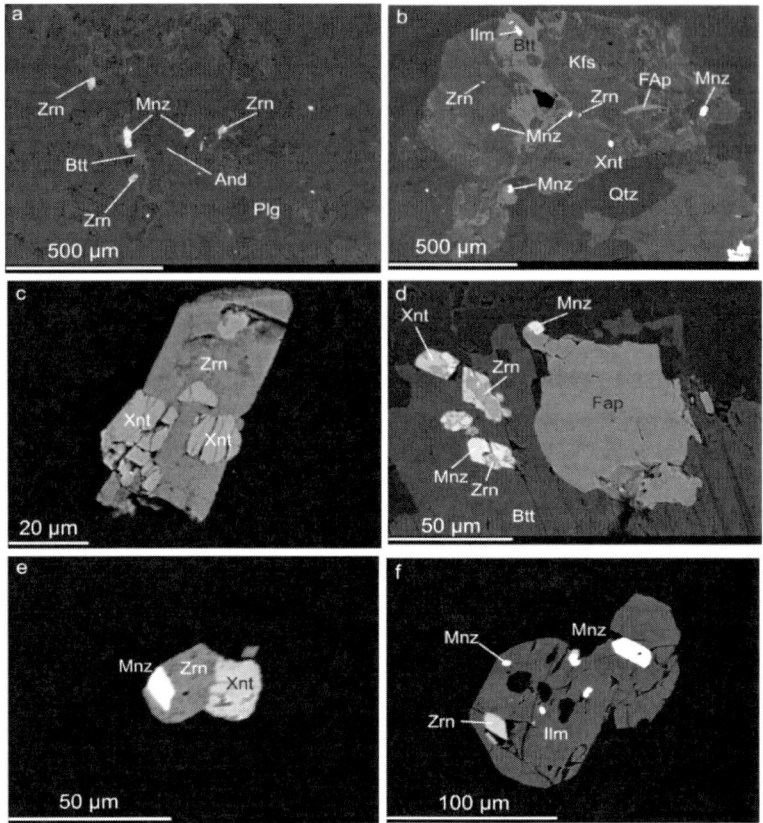

Figure 8. High contrast BSE images of accessory minerals from two-mica granites of the Deštná type (South Bohemian Batholith). Mineral abbreviations according to Whitney and Evans [56].

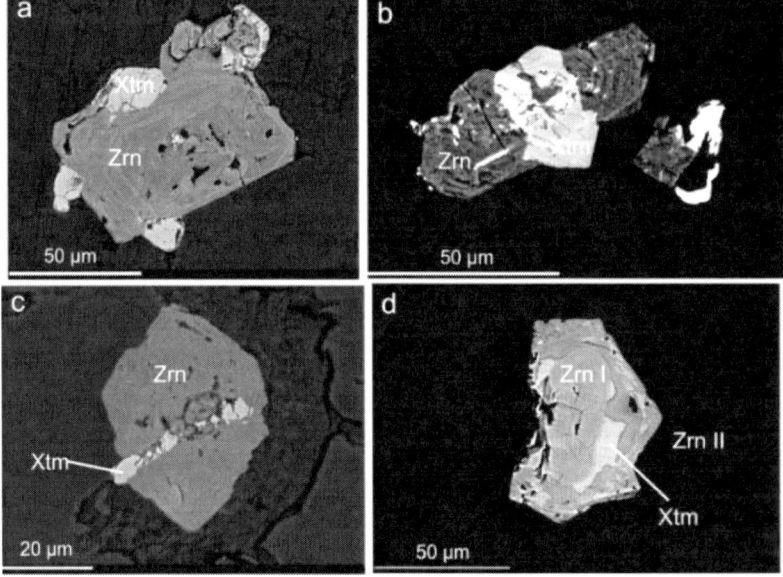

Figure 9. High contrast BSE images of zircon and xenotime from P-rich granites of the Horní Slavkov–Krásno ore district. Mineral abbreviations according to Whitney and Evans [56].

## Zircon and Xenotime in Topaz Granites

Zircon in both topaz granite suites is of variable size (10–60 μm), subhedral to euhedral. Most zircon grains are included in lithium mica. Zircon very often shows strong hydration and fluorination. The amount of molecular water was estimated from the analytical deficit according to Johan and Johan [7]. Xenotime in both granite suites forms intergrowths with zircon. In the P-rich granites from Hub stock (Horní Slavkov–Krásno area) xenotime grains are associated with zircon either as grains on the zircon rim (Figure 9 a) or inclusions in bigger zircon crystals (Figure 9 b, c). Very rare are more complex intergrowths between both minerals with a younger zircon rim (Figure 9 d). In the P-poor granite from the Cínovec cupola zircon–xenotime intergrowths are more complex. As well as in the P-rich granite xenotime forms younger veins in altered zircon crystals (Figure 10 a). However, more abundant are epitaxial overgrowths of xenotime on zircon coupled with their inclusions (exsolutions?) in altered zircon grains (Figure 10 b, c). In some cases xenotime forms bigger subhedral grains enclosed in fluorite with inclusions of zircon, REE hydroxyfluorides and synchysite-(Y) (Figure 10 d).

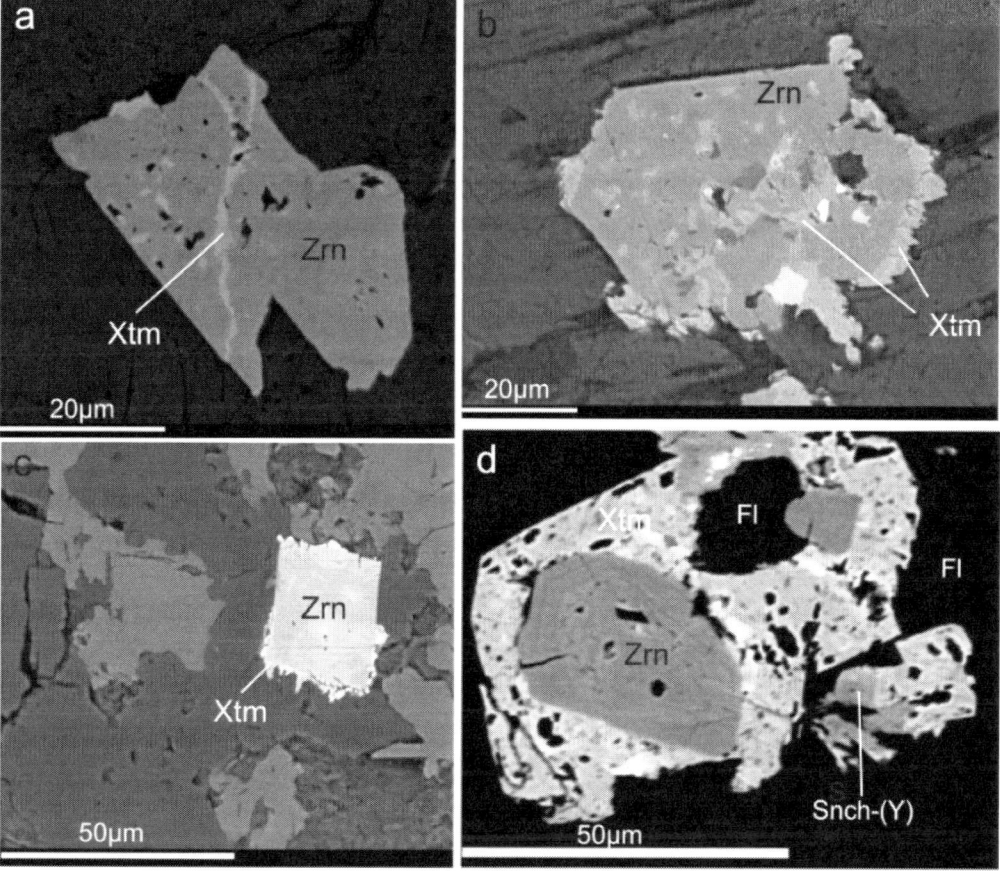

Figure 10. High contrast BSE images of zircon and xenotime from P-poor granites of the Cínovec cupola. Mineral abbreviations according to Whitney and Evans [56].

## COMPOSITION OF ACCESSORY MINERALS

### Monazite

Composition of monazite was studied only in two-mica granites of the SBB. Monazite forms a part of an isostructural series of monoclinic minerals with the general formula $ABO_4$, where A = REE, Y, Th, U, Ca, Pb and B = P, Si. Because La, Ce, and Nd are typically the dominant LREE in natural monazites, we follow the proposal of the IMA in distinguishing the members monazite-(La), monazite-(Ce), and monazite-(Nd) [4]. Huttonite ($ThSiO_4$) and cheralite [$CaTh(PO_4)_2$] are the other prominent end-members [37, 38, 39]. Abundances of $ThO_2$ in monazite from S-type granites typically range from 4 to 12 wt.% [1, 2, 3, 4, 40, 41, 42, 43].

The monazite in two-mica granites of the Eisgarn ad Deštná types is Ce-dominant, with concentrations of Ce in range between 1.28 and 1.76 apfu Ce (Table 1). The amounts of other LREE are distinctly lower (La 0.52–0.94 apfu, Nd 0.56–0.80 apfu, Sm 0.08–0.19 apfu). Typical concentrations of HREE in monazites from both granite types range between 0.04 and 0.22 apfu (Gd 0.03–0.14 apfu, Dy 0.00–0.08 apfu). Amounts of xenotime ($YPO_4$) component are somewhat higher in monazites from the Deštná granite (up to 8.6 mol.%). In monazites from the Eisgarn granite amounts of xenotime component range from 0.4 to 5.6 mol.% of $YPO_4$. The huttonite to cheralite ratio is relatively constant with increasing (Th + U) abundances. The amounts of huttonite ($ThSiO_4$) component in monazite-(Ce) from the Eisgarn granite ranges between zero and 5.0 mol. %. However, in monazite-(Ce) from the Deštná granite this content is distinctly lower (up to 1.0 mol. %). Similar evolution displays content of cheralite [Ca, Th ($PO_4)_2$] component; 7.3–23.3 mol.% in monazites from the Eisgarn granite and 8.8–12.6 mol.% in monazites from the Deštná granite (Figure 11 a).

Analyzed monazites from both granite types typically show preference for Th over U, resulting in higher Th/U ratio. However, values of this ratio in monazites from both granite types are highly variable; from 1.5 to 40.0 in monazites of the Eisgarn granite and from 1.3 to 15.4 in monazites of the Deštná granite (Figure 11 b).

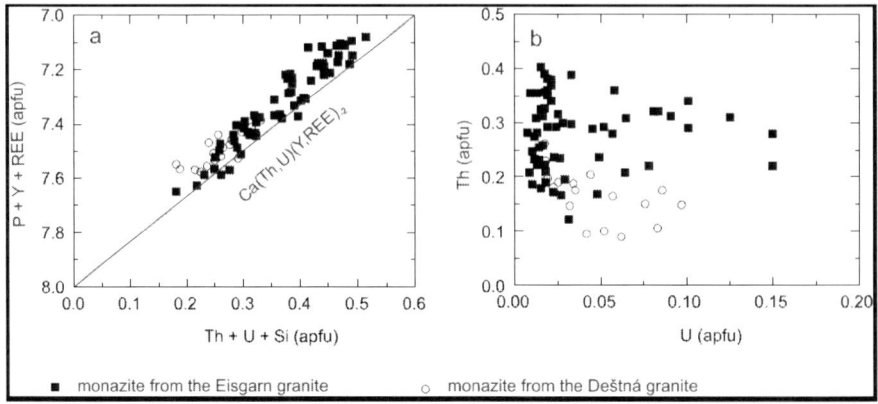

Figure 11 a) Plot of P + Y + REE vs. Th + U + Si for monazite from two-mica granites of the SBB, b) Plot of Th vs. U for monazite from two-mica granites of the SBB.

Table 1. Representative analysis of monazites from two-mica granites of the SBB

| wt.% | 1307-35 Eisgarn | 1307-43 Eisgarn | 1310-7 Eisgarn | 1310-11 Eisgarn | 735-14 Deštná | 1486-15 Deštná | 1486-23 Deštná | 1487-4 Deštná |
|---|---|---|---|---|---|---|---|---|
| $P_2O_5$ | 30.57 | 29.79 | 29.77 | 28.83 | 29.65 | 30.05 | 30.01 | 30.04 |
| $SiO_2$ | 0.22 | 0.38 | 0.30 | 0.31 | 0.32 | 0.34 | 0.22 | 0.28 |
| $ThO_2$ | 9.10 | 8.70 | 9.56 | 5.77 | 6.40 | 4.96 | 4.20 | 5.55 |
| $UO_2$ | 2.44 | 0.47 | 2.90 | 0.51 | 0.52 | 2.46 | 2.80 | 0.55 |
| $Y_2O_3$ | 2.56 | 0.67 | 2.55 | 0.85 | 1.36 | 2.21 | 3.11 | 1.19 |
| $La_2O_3$ | 9.64 | 11.85 | 8.94 | 13.62 | 13.61 | 11.27 | 12.03 | 13.58 |
| $Ce_2O_3$ | 23.26 | 26.80 | 22.32 | 29.71 | 27.62 | 26.21 | 25.86 | 28.34 |
| $Pr_2O_3$ | 2.77 | 3.06 | 2.80 | 3.24 | 2.91 | 3.01 | 2.80 | 3.12 |
| $Nd_2O_3$ | 10.19 | 11.29 | 10.45 | 11.82 | 10.43 | 11.18 | 10.58 | 11.63 |
| $Sm_2O_3$ | 2.59 | 2.11 | 3.45 | 2.40 | 1.87 | 2.81 | 2.49 | 1.93 |
| $Gd_2O_3$ | 1.90 | 1.26 | 2.67 | 1.25 | 1.44 | 1.90 | 1.86 | 1.23 |
| $Dy_2O_3$ | 1.15 | 0.27 | 1.22 | 0.39 | 0.59 | 1.01 | 1.21 | 0.54 |
| $Er_2O_3$ | 0.16 | 0.09 | 0.16 | 0.06 | 0.06 | 0.13 | 0.19 | 0.08 |
| CaO | 2.45 | 1.68 | 2.47 | 1.11 | 1.32 | 1.47 | 1.48 | 1.18 |
| FeO | 0.06 | 0.62 | 0.00 | 0.02 | 0.33 | 0.04 | 0.02 | 0.00 |
| PbO | 0.26 | 0.15 | 0.29 | 0.10 | 0.11 | 0.18 | 0.20 | 0.11 |
| $Sc_2O_3$ | 0.00 | 0.00 | 0.00 | 0.00 | 0.01 | 0.01 | 0.00 | 0.00 |
| $Al_2O_3$ | 0.00 | 0.00 | 0.00 | 0.01 | 0.01 | 0.00 | 0.01 | 0.00 |
| $As_2O_5$ | 0.05 | 0.06 | 0.00 | 0.00 | 0.04 | 0.06 | 0.07 | 0.06 |
| Total | 99.37 | 99.25 | 99.85 | 100.02 | 98.60 | 99.30 | 99.14 | 99.41 |
| apfu (O=8) | | | | | | | | |
| P | 4.009 | 3.963 | 3.949 | 3.887 | 3.964 | 3.973 | 3.971 | 3.982 |
| Si | 0.034 | 0.060 | 0.047 | 0.049 | 0.051 | 0.053 | 0.034 | 0.044 |
| Th | 0.321 | 0.311 | 0.341 | 0.209 | 0.230 | 0.176 | 0.149 | 0.198 |
| U | 0.084 | 0.016 | 0.101 | 0.018 | 0.018 | 0.086 | 0.097 | 0.019 |
| Y | 0.211 | 0.056 | 0.212 | 0.072 | 0.114 | 0.183 | 0.258 | 0.099 |
| La | 0.550 | 0.686 | 0.516 | 0.799 | 0.792 | 0.649 | 0.693 | 0.783 |
| Ce | 1.318 | 1.540 | 1.279 | 1.730 | 1.595 | 1.497 | 1.478 | 1.623 |
| Pr | 0.156 | 0.175 | 0.160 | 0.188 | 0.167 | 0.171 | 0.159 | 0.178 |
| Nd | 0.563 | 0.633 | 0.584 | 0.672 | 0.588 | 0.623 | 0.590 | 0.650 |
| Sm | 0.138 | 0.114 | 0.186 | 0.132 | 0.102 | 0.151 | 0.134 | 0.104 |

**Table 1. (Continued)**

|    | 1307-35 | 1307-43 | 1310-7 | 1310-11 | 735-14 | 1486-15 | 1486-23 | 1487-4 |
|----|---------|---------|--------|---------|--------|---------|---------|--------|
| Gd | 0.097   | 0.066   | 0.139  | 0.066   | 0.075  | 0.098   | 0.096   | 0.064  |
| Dy | 0.057   | 0.014   | 0.062  | 0.020   | 0.030  | 0.051   | 0.061   | 0.027  |
| Er | 0.008   | 0.004   | 0.008  | 0.003   | 0.003  | 0.006   | 0.009   | 0.004  |
| Ca | 0.407   | 0.283   | 0.198  | 0.189   | 0.223  | 0.246   | 0.248   | 0.198  |
| Fe | 0.008   | 0.081   | 0.000  | 0.003   | 0.044  | 0.005   | 0.003   | 0.000  |
| Pb | 0.011   | 0.006   | 0.012  | 0.004   | 0.005  | 0.008   | 0.008   | 0.005  |
| Sc | 0.000   | 0.000   | 0.000  | 0.000   | 0.001  | 0.001   | 0.001   | 0.000  |
| Al | 0.000   | 0.000   | 0.000  | 0.000   | 0.002  | 0.000   | 0.000   | 0.000  |
| As | 0.004   | 0.005   | 0.000  | 0.000   | 0.003  | 0.005   | 0.006   | 0.005  |

## Xenotime

Xenotime-(Y) is particularly abundant in Ca-poor peraluminous granites accounting for a large fraction of Y and HREE contents in whole rocks [3, 5, 44]. Xenotimes-(Y) from these granitic rocks always contain 5–15 wt.% of HREE [3]. However, xenotime-(Y) from weakly peraluminous, Y-enriched granite from Xihuashan, southern China contains up to 25.2 wt.% $HREE_2O_3$ [45]. On the other hand, xenotime-(Y) from A-type Laoshan granite (Eastern China) contains very low concentrations of HREE (2.7–5.8 wt.% $HREE_2O_3$) [46]. Two main mechanisms for the replacement of Y by REE, U and Th exists, namely charge-balancing coupled substitution involving Si and Ca (xenotime-type and/or thorite-coffinite-type substitution and brabantite-type substitution). The more dominant thorite (coffinite) solid solution, represented by a vector $(Th,U)SiREE_{-1}P_{-1}$ corresponds to the cationic substitution $(REE, Y)^{3+} + P^{5+} \Leftrightarrow (Th, U)^{4+} + Si^{4+}$ [5, 44]. The brabantite-type substitution, represented by a vector $(U,Th)CaREE_{-2}$ corresponds to substitution $2 (REE,Y)^{3+} \Leftrightarrow (U,Th)^{4+} + Ca^{2+}$ [5]. Representative results of electron probe analyses of xenotime from investigated two-mica- and topaz granites are listed in Table 2.

In the two-mica granites of the SBB, xenotime-(Y) was only found in the Deštná granite. The proportion of $YPO_4$, the main component in xenotime, ranges in xenotime from the Deštná granite from 71.4 to 77.1 mol.%. The HREE content in examined xenotime ranges from 0.73 to 0.96 apfu (18.2–21.7 mol.% $HREEPO_4$. The concentration ranges for the individual HREE oxides are as follows (in apfu): Gd 0.07–0.20, Tb 0.02–0.05, Dy 0.23–0.37, Ho 0.04–0.07, Er 0.13–0.18, Yb 0.02–0.24, Lu 0.01–0.05. Concentrations of LREE are typically low compared to the HREE in xenotime. Concentrations of Ce range from below the detection limit of the electron microprobe to 0.01 apfu. Concentrations of Nd and Sm range from 0.02 to 0.04 apfu Nd and 0.02 to 0.07 apfu Sm. Examined xenotime grains have Sm/Nd ratios between 1.1 and 2.1. Total LREE content in examined xenotimes range from 1.1 to 2.8 mol.% $LREEPO_4$.

The data points for xenotime grains from the Deštná granite are shifted toward the vector Ca (U, Th) (Y, REE)$_{-2}$ (Figure 12a). The examined xenotime incorporates distinctly higher concentrations of U than Th and displays Th/U = 0.08–0.33 (Figure 12 b).

In topaz granites of the Krušné Hory/Erzgebirge area, xenotime-(Y) occurs in both granite suites, but is more abundant in P-poor granites from the Cínovec cupola, especially in protolithionite granite. Microprobe analyses reveal low totals suggesting significant hydration of xenotime. The water content estimated from analytical data ranges from 5 to 11 wt.% $H_2O$ in xenotime from the P-rich granites (Horní Slavkov–Krásno area) and from 1 to 12 wt.% $H_2O$ in xenotime from the P-poor granites of the Cínovec cupola.

The F content in xenotime is similar to its content in zircon, in slightly greisenized granite from the Hub stock it reaches up to 1.12 wt.%. However, xenotime from P-poor granites of the Cínovec contains a somewhat higher amount of F (up to 2.4 wt.%). Xenotime, occurring in granitic rocks, characterized by the ideal formula of $YPO4$, is commonly enriched in HREE, Th and U. The HREE content is distinctly high in xenotime from analyzed P-poor granite (12.2–35.7 wt.% $HREE_2O_3$). In these xenotimes concentrations of Nd and Sm range from 0.01 to 0.79 wt.% $Nd_2O_3$ (0.001–0.049 apfu Nd) and 0.01 to 1.60 wt.% $Sm_2O_3$ (0.001–0.095 apfu Sm).

Table 2. Representative chemical analyses of xenotime

| wt.% | 588-18 | 1486-24 | 1007-4 | 1007-11 | 1007-19 | 629-1 | 629-6 | 633-11 |
|---|---|---|---|---|---|---|---|---|
| | Deštná | Deštná | Krásno | Krásno | Krásno | Cínovec | Cínovec | Cínovec |
| $P_2O_5$ | 34.82 | 32.66 | 34.50 | 36.31 | 33.87 | 23.92 | 22.39 | 22.57 |
| $SiO_2$ | 0.16 | 0.34 | 0.96 | 0.00 | 0.85 | 0.95 | 1.02 | 0.55 |
| $ThO_2$ | 0.10 | 0.28 | 0.87 | 0.04 | 0.72 | 2.77 | 1.31 | 0.24 |
| $UO_2$ | 0.99 | 1.99 | 4.43 | 0.93 | 4.56 | 1.27 | 1.44 | 0.07 |
| $Y_2O_3$ | 42.34 | 42.14 | 38.83 | 48.50 | 38.23 | 25.26 | 23.27 | 42.00 |
| $La_2O_3$ | 0.00 | 0.00 | 0.00 | 0.03 | 0.01 | 0.03 | 0.01 | 0.02 |
| $Ce_2O_3$ | 0.11 | 0.13 | 0.00 | 0.00 | 0.03 | 0.15 | 0.10 | 0.00 |
| $Pr_2O_3$ | 0.12 | 0.00 | 0.00 | 0.02 | 0.07 | 0.08 | 0.06 | 0.00 |
| $Nd_2O_3$ | 0.32 | 0.56 | 0.32 | 0.08 | 0.42 | 0.79 | 0.24 | 0.03 |
| $Sm_2O_3$ | 0.66 | 0.81 | 0.87 | 0.11 | 0.79 | 1.60 | 0.85 | 0.13 |
| $Gd_2O_3$ | 2.58 | 2.58 | 2.56 | 0.61 | 2.38 | 3.30 | 2.18 | 0.91 |
| $Tb_2O_3$ | 0.72 | 0.66 | 0.00 | 0.00 | 0.00 | 0.77 | 0.48 | 0.67 |
| $Dy_2O_3$ | 6.22 | 6.16 | 7.59 | 3.79 | 7.35 | 9.36 | 8.41 | 5.20 |
| $Ho_2O_3$ | 1.10 | 1.23 | 0.98 | 0.69 | 1.06 | 2.07 | 2.69 | 1.73 |
| $Er_2O_3$ | 3.41 | 3.46 | 3.23 | 2.48 | 3.42 | 6.26 | 7.27 | 4.81 |
| $Tm_2O_3$ | 0.57 | 0.61 | 0.00 | 0.00 | 0.00 | 1.27 | 1.93 | 0.87 |
| $Yb_2O_3$ | 3.07 | 3.00 | 4.23 | 3.81 | 3.96 | 10.88 | 14.69 | 7.25 |
| $Lu_2O_3$ | 0.62 | 0.53 | 0.81 | 0.56 | 0.91 | 1.46 | 2.68 | 1.35 |
| CaO | 0.10 | 0.30 | 0.42 | 0.89 | 0.51 | 0.09 | 0.04 | 0.12 |
| PbO | 0.38 | 0.40 | 0.18 | 0.02 | 0.10 | 0.26 | 0.26 | 0.28 |
| $Sc_2O_3$ | 0.04 | 0.05 | 0.00 | 0.00 | 0.00 | 0.05 | 0.17 | 0.16 |
| $Bi_2O_3$ | 0.00 | 0.00 | 0.00 | 0.00 | 0.00 | 0.00 | 0.03 | 0.00 |
| $Nb_2O_5$ | 0.00 | 0.00 | 0.00 | 0.00 | 0.00 | 0.03 | 0.04 | 0.07 |
| $WO_3$ | 0.00 | 0.00 | 0.00 | 0.00 | 0.00 | 0.08 | 0.13 | 0.16 |
| $ZrO_2$ | 0.02 | 0.05 | 0.00 | 0.00 | 0.00 | 0.02 | 0.53 | 1.83 |
| F | 0.00 | 0.00 | 0.09 | 1.11 | 0.10 | 0.45 | 0.32 | 0.77 |
| O=F | 0.00 | 0.00 | 0.04 | 0.47 | 0.04 | 0.19 | 0.13 | |
| Total | 98.45 | 97.94 | 100.83 | 99.51 | 99.30 | 92.98 | 92.41 | 90.25 |

| apfu (O=8) | 588-18 | 1486-24 | 1007-4 | 1007-11 | 1007-19 | 629-1 | 629-6 | 633-11 |
|---|---|---|---|---|---|---|---|---|
| P | 3.973 | 3.841 | 3.939 | 4.008 | 3.935 | 3.485 | 3.376 | 3.253 |
| Si | 0.022 | 0.047 | 0.129 | 0.000 | 0.117 | 0.163 | 0.182 | 0.094 |
| Th | 0.003 | 0.009 | 0.027 | 0.001 | 0.022 | 0.108 | 0.053 | 0.009 |
| U | 0.030 | 0.062 | 0.133 | 0.027 | 0.139 | 0.049 | 0.057 | 0.003 |
| Y | 3.034 | 3.112 | 2.784 | 3.362 | 2.789 | 2.311 | 2.203 | 3.801 |
| La | 0.000 | 0.000 | 0.000 | 0.001 | 0.001 | 0.002 | 0.001 | 0.001 |
| Ce | 0.005 | 0.007 | 0.000 | 0.000 | 0.002 | 0.009 | 0.007 | 0.000 |
| Pr | 0.000 | 0.000 | 0.000 | 0.001 | 0.003 | 0.005 | 0.004 | 0.000 |
| Nd | 0.015 | 0.028 | 0.015 | 0.004 | 0.021 | 0.049 | 0.015 | 0.002 |
| Sm | 0.031 | 0.039 | 0.040 | 0.005 | 0.037 | 0.095 | 0.052 | 0.008 |
| Gd | 0.115 | 0.119 | 0.114 | 0.026 | 0.108 | 0.188 | 0.129 | 0.051 |
| Tb | 0.032 | 0.030 | 0.000 | 0.000 | 0.000 | 0.043 | 0.028 | 0.037 |
| Dy | 0.270 | 0.275 | 0.329 | 0.159 | 0.325 | 0.518 | 0.482 | 0.285 |
| Ho | 0.047 | 0.054 | 0.042 | 0.029 | 0.046 | 0.113 | 0.152 | 0.094 |
| Er | 0.144 | 0.151 | 0.137 | 0.101 | 0.147 | 0.338 | 0.406 | 0.257 |
| Tm | 0.024 | 0.026 | 0.000 | 0.000 | 0.000 | 0.068 | 0.107 | 0.046 |
| Yb | 0.126 | 0.127 | 0.174 | 0.151 | 0.166 | 0.571 | 0.797 | 0.376 |
| Lu | 0.025 | 0.022 | 0.033 | 0.022 | 0.038 | 0.076 | 0.144 | 0.069 |
| Ca | 0.014 | 0.045 | 0.061 | 0.124 | 0.075 | 0.017 | 0.008 | 0.022 |
| Pb | 0.014 | 0.015 | 0.007 | 0.001 | 0.004 | 0.012 | 0.012 | 0.013 |
| Sc | 0.005 | 0.006 | 0.000 | 0.000 | 0.000 | 0.007 | 0.026 | 0.024 |
| Bi | 0.000 | 0.000 | 0.000 | 0.000 | 0.000 | 0.000 | 0.001 | 0.000 |
| Nb | 0.000 | 0.000 | 0.000 | 0.000 | 0.000 | 0.002 | 0.003 | 0.000 |
| W | 0.000 | 0.000 | 0.000 | 0.000 | 0.000 | 0.004 | 0.006 | 0.003 |
| Zr | 0.001 | 0.003 | 0.000 | 0.000 | 0.000 | 0.002 | 0.046 | 0.013 |
| F | 0.000 | 0.000 | 0.077 | 0.924 | 0.087 | 0.490 | 0.360 | 1.970 |

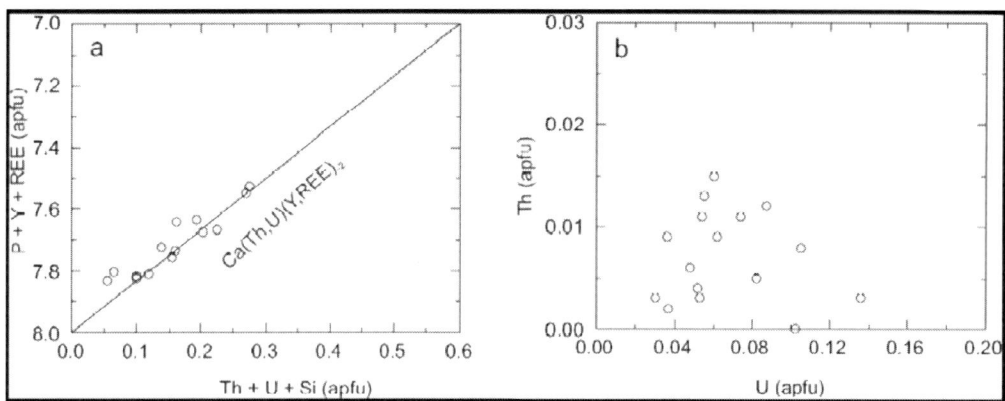

Figure 12. a) Plot of P + Y + REE vs. Th + U + Si for xenotime from two-mica granites of the SBB, b) Plot of Th vs. U for xenotime from from two-mica granites of the SBB.

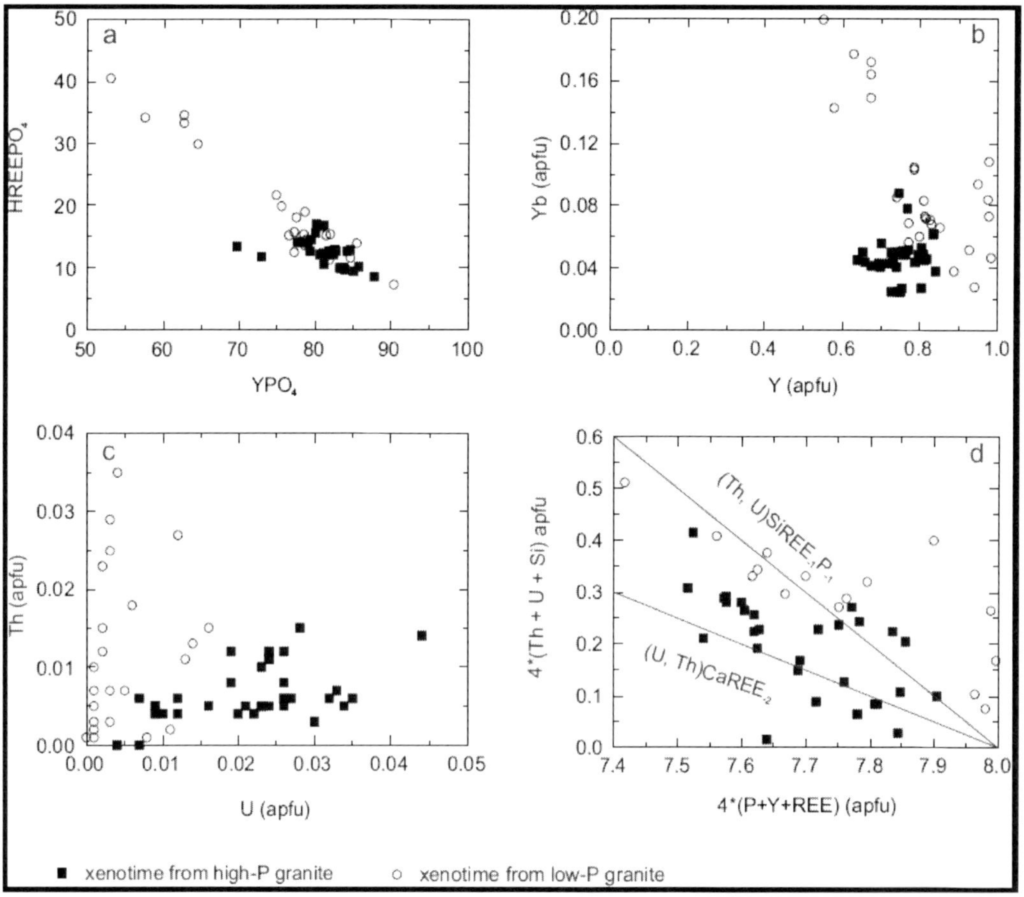

Figure 13. a) Plot HREEPO$_4$ vs. YPO$_4$ (mol. %) for xenotime from P-poor and P-rich granites of the Krušné Hory/Erzgebirge area, b) Plot Yb vs. Y for xenotime from P-poor and P-rich granites of the Krušné Hory/Erzgebirge area, c) Plot Th vs. U for xenotime from P-poor and P-rich granites of the Krušné Hory/Erzgebirge area, d) Plot 4* (Th + U + Si) vs. 4* (P + Y + REE) for xenotime from P-poor and P-rich granites of the Krušné Hory/Erzgebirge area.

Xenotimes from P-rich granites contain considerably lower HREE concentrations (9.3–19.5 wt.% $HREE_2O_3$). In xenotimes from P-rich granite, the concentrations of Nd and Sm range from 0.04 to 0.56 wt.% $Nd_2O_3$ (0.001–0.029 apfu Nd) and 0.11 to 0.87 wt.% $Sm_2O_3$ (0.001–0.043 apfu Sm). The proportion of $YPO_4$, the main component in xenotime, ranges from 53.0.0 to 90.3 mol.% in P-poor granite and from 64.9 to 87.9 mol.% in P-rich granites (Figure 13 a). The concentrations of Dy and Yb range from 2.39 to 9.36 wt.% $Dy_2O_3$ (0.11–0.52 apfu Dy) and 2.48 to 14.69 wt.% $Yb_2O_3$ (0.11–0.80 apfu Yb) in xenotimes from P-poor granite. In xenotimes from P-rich granite the concentrations of Dy and Yb range from 3.05 to 7.67 wt.% $Dy_2O_3$ (0.14–0.34 apfu Dy) and 2.24 to 8.04 wt.% $Yb_2O_3$ (0.10–0.35 apfu Yb) (Figure 13 b).

The $HREEPO_4$ component ranges from 8.6 to 17.0 mol.% xenotime in P-rich granites. However, concentrations of this component in xenotime from P-poor granites are distinctly higher (7.4–40.5 mol.%). Concentrations of U and Th range from 0.45 to 5.6 wt.% $UO_2$ (0.01–0.17 apfu U) and 0.04–1.9 wt. % $ThO_2$ (0.001–0.14 apfu Th) in xenotimes from P-rich granite. In xenotimes from P-poor granite the U/Th ratio is inversed with 0.04–1.84 wt. % $UO_2$ (0.001–0.07 apfu U) and 0.08–4.23 wt.% $ThO_2$ (0.003–0.14 apfu Th) (Figure 13 c). Two main mechanisms for the replacement of Y by REE, U and Th exists, namely charge-balancing coupled substitution involving Si and Ca (xenotime-type and/or thorite-coffinite-type substitution and brabantite-type substitution). In xenotime from the studied topaz granites both substitution mechanisms occur. For xenotime from the P-rich granite the brabantite-type substitution is more significant, whereas for xenotime from the P-poor granite is thorite-type substitution significant (Figure 13 d). The brabantite-type substitution was also found in S-type topaz granites from the German part of the Krušné Hory/Erzgebirge area [5]. The importance of thorite-type substitution for P-poor granites from this area is also supported by occurrence of Th-rich xenotime an/or Y-rich thorite in these granites [47].

For altered xenotime from both topaz granite suites there is also significant enrichment in As, Nb, W, Sc and Bi. The distribution of these five elements in xenotime from investigated granites is heterogeneous at every scale of observation. Xenotime grains enriched in these elements occur together with xenotime containing As, Nb, W, Sc and Bi below the microprobe detection limit. The highest As concentrations were found in xenotime from altered P-poor, zinnwaldite-bearing granite occurring in upper part of the Cínovec cupola (up to 0.42 wt.% $As_2O_5$; 0.04 apfu As). Xenotime from P-poor granite is also enriched in W (up to 0.22 wt.% $WO_3$; 0.01 apfu W), whereas xenotime from P-rich granite is in rare cases enriched in Sc (up to 2.03 wt.% $Sc_2O_3$; 0.25 apfu Sc) and in Bi (up to 0.07 wt.% $Bi_2O_3$).

## Zircon

Although several minerals are isostructural with zircon (e.g., xenotime, thorite and coffinite), only hafnon ($HfSiO_4$) displays complete solid solution. Granitic zircon is typically enriched in Hf, especially in highly fractionated granites [48, 49, 50, 51] with $HfO_2$ abundances up to 35 wt.% [50]. The Zr/Hf ratio of the zircon is usually regarded as an important indicator of granite melt fractionation level [49, 51, 52]. Yttrium variation is coupled to P content and multi-element correlation implies that the isomorphic coupled substitution $(REE + Y)^{3+} + P^{5+} \Leftrightarrow Zr^{4+} + Si^{4+}$ (xenotime-type) is operating in zircon. This substitution has been predominantly suggested for zircons from topaz granites [7, 51]. In the two-mica granites abundances of Y and P are restricted, usually ranging from 0.01 to 0.07 apfu Y [43, 52, 53, 54, 55]. Furthermore in some zircon a positive correlation between U and Th is observed [54, 55], which could be explained by a combination of thorite-type and coffinite-type substitutions.

Representative results of electronprobe analyses of zircon from investigated two-mica- and topaz granites are listed in Table 3. Analyzed zircons from both two-mica granite types of the SBB contain rather low Hf concentrations (1.0–2.5 wt.% $HfO_2$, equivalent to 0.009–0.02 Hf apfu). The proportion of the hafnium end member indicated by atom ratio Hf/(Zr + Hf) varies from 0.01 to 0.02 in both granite types (Figure 14 a). The relatively low enrichment in P is associated with a simultaneous enrichment in (Y+REE) and occurs in both granite types. However, enrichment of P, Y and REE is greater in the Deštná granite (up to 0.05 apfu P, 0.03 apfu Y and 0.01 apfu REE) (Figure 14 b).

The LREE abundances are mostly below their detection limits of microprobe. The contents of the HREE are variable at all scales and range from below their detection limits to 0.006 apfu in the Eisgarn granite and to 0.01 in the Deštná granite. The U concentrations of U in both granite types are also variable and range from below its detection limit to 0.03 apfu.

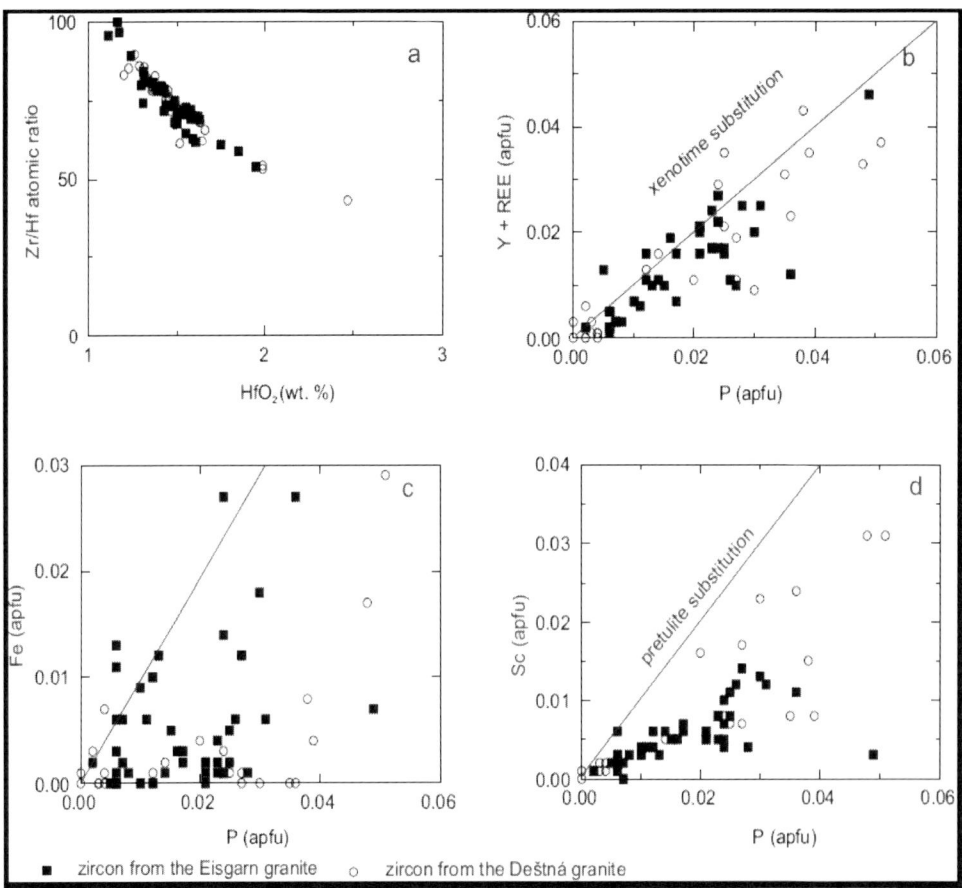

Figure 14. a) Plot of Zr/Hf vs. $HfO_2$ for zircon from two mica granites of the SBB, b)Plot Y + REE vs. P for zircon from two-mica granites of the SBB, c) Plot Fe vs. P for zircon from two-mica granites of the SBB, d) Plot Sc vs. P for zircon from two-mica granites of the SBB.

Its highest contents were found in altered zircons from the Eisgarn granite. In these cases uranium displays a positive correlation with calcium and fluorine. The content of Th is distinctly lower and frequently it is below the detection limit. The highest concentration of Th was found in the Eisgarn granite (up to 0.007 apfu Th). It is significant that zircons from the Eisgarn granite are enriched in Fe (up to 0.03 apfu Fe), which displays a positive correlation with P (Figure 14 c). However, zircons from the Deštná granite are partly enriched in Sc (up to 0.03 apfu Sc) (Figure 14 d). The concentrations of Bi, As, Nb and W in investigated zircons from both granite types are usually below the microprobe detection limit. Zircon from topaz granites of the Krušné Hory/Erzgebirge area are often hydrated and fluorized. The highest concentrations of $H_2O$ and F are displayed by zircon crystals from the apical part of the Cínovec cupola (max. 18.5 wt.% $H_2O$ and 0.1–1.7 wt.% F). The hydration and fluorination of zircon from the Krásno–Horní Slavkov ore district is likewise high. Fluorine reaches 1.6 wt. % in zircon from partly greisenized topaz-albite granite of the Hub stock. All analyzed zircons from both topaz granite suites contain rather moderate Hf concentrations (1.2–8.2 wt.% $HfO_2$, equivalent to 0.012–0.073 Hf apfu). The proportion of the hafnium end member indicated by atomic ratio Hf/(Zr + Hf) varies from 0.01 to 0.09 (P-rich granite, Hub stock 0.01–0.08, P-poor granite, Cínovec 0.01–0.09) (Figure 15 a). In zircon from P-rich granites of the Hub stock enrichment in P, which is not associated with a simultaneous enrichment in (Y + REE) (Figure 15 b), was observed. This high concentration of P (up to 8.29 wt.% $P_2O_5$, equivalent to 0.24 apfu P) is typically accompanied by elevated enrichment of Ca and Al. The concentrations of P in zircons from P-poor granite of the Cínovec cupola range between 0.2 and 2.6 wt.% $P_2O_5$ (0.006–0.08 apfu P). Aluminum is positively correlated with P. Zircon from P-rich granites of the Horní Slavkov–Krásno ore district containing 0.02 to 2.0 wt.% $Al_2O_3$ (0.001–0.08 apfu Al). The concentration of Al in zircon from P-poor granites of the Cínovec cupola is somewhat lower (0.1–1.5 wt.% $Al_2O_3$; 0.004–0.06 apfu Al) (Figure 15 c). Calcium usually displays a positive correlation with uranium. U-rich zircon from P-rich granites contains up to 5.4 wt.% CaO (0.20 apfu Ca). However, zircon from P-poor granites contains up to 1.6 wt.% CaO (0.06 apfu Ca).

The distribution of iron is heterogeneous. Measured Fe abundances range from 0.01 to 4.3 wt.% FeO (0–0.13 apfu Fe) in zircons from P-rich granites and from 0.12 to 2.3 wt.% FeO (0.004–0.07 apfu Fe) in zircons from P-poor granites.

Zircon from both topaz granite suites is also partly enriched in Sc. The highest amount of Sc was found in zircon from P-rich granites of the Hub stock (1.17 wt.% $Sc_2O_3$; 0.03 apfu Sc). In zircon from P-poor granites of the Cínovec cupola the highest measured Sc concentration was 0.42 wt.% $Sc_2O_3$ (0.014 apfu Sc) (Figure 15 d). Zircons from analyzed P-rich and P-poor topaz granites show significant differences in the contents of Y. Zircons from the P-poor granites displays a higher enrichment in Y (up to 18.4 wt.% $Y_2O_3$; 0.34 apfu Y), whereas Y concentrations in P-rich granites reach only 5.5 wt.% $Y_2O_3$ (0.10 apfu Y). The LREE abundances are mostly below their detection limits of the microprobe. The contents of the HREE are variable at all scales and range from below their detection limits to 8.6 wt.% $HREE_2O_3$. The higher enrichment in HREE displays zircon from P-poor granites (0.4–8.6 wt.% $HREE_2O_3$; 0.004–0.10 apfu HREE). The zircon from P-rich granites displays considerably lower amounts of HREE (0.1–1.2 wt.% $HREE_2O_3$; 0.001–0.01 apfu HREE).

**Table 3. Representative chemical analyses of zircon**

| Wt. % | 1307-44 | 1310-14 | 735-5 | 1487-3 | 1283-22 | 1441-26 | 633-6 | 634-37 |
|---|---|---|---|---|---|---|---|---|
| | Eisgarn | Eisgarn | Deštná | Deštná | Krásno | Krásno | Cínovec | Cínovec |
| SiO$_2$ | 31.55 | 31.78 | 31.54 | 31.73 | 29.08 | 22.47 | 28.13 | 30.69 |
| Al$_2$O$_3$ | 0.00 | 0.00 | 0.06 | 0.00 | 0.37 | 1.13 | 0.15 | 0.16 |
| ZrO$_2$ | 63.38 | 64.63 | 62.12 | 64.10 | 56.23 | 50.94 | 56.51 | 61.03 |
| HfO$_2$ | 1.49 | 1.34 | 1.36 | 1.44 | 2.34 | 2.60 | 1.50 | 2.71 |
| CaO | 0.08 | 0.00 | 0.01 | 0.02 | 0.31 | 3.59 | 0.71 | 0.23 |
| FeO | 0.09 | 0.19 | 0.09 | 0.08 | 0.00 | 1.49 | 1.42 | 1.03 |
| MnO | 0.00 | 0.04 | 0.00 | 0.01 | 0.02 | 0.18 | 0.38 | 0.22 |
| MgO | 0.01 | 0.00 | 0.01 | 0.02 | 0.01 | 0.08 | 0.03 | 0.00 |
| P$_2$O$_5$ | 0.88 | 0.58 | 0.94 | 0.55 | 1.49 | 3.78 | 0.41 | 0.63 |
| Sc$_2$O$_3$ | 0.18 | 0.17 | 0.40 | 0.20 | 0.47 | 0.59 | 0.02 | 0.01 |
| As$_2$O$_5$ | 0.04 | 0.00 | 0.00 | 0.00 | 0.07 | 0.35 | 0.03 | 0.00 |
| Bi$_2$O$_3$ | 0.07 | 0.06 | 0.08 | 0.11 | 0.09 | 0.07 | 0.08 | 0.03 |
| Nb$_2$O$_5$ | 0.00 | 0.00 | 0.00 | 0.00 | 0.00 | 0.22 | 0.02 | 0.00 |
| WO$_3$ | 0.00 | 0.00 | 0.00 | 0.00 | 0.00 | 0.63 | 0.05 | 0.12 |
| Y$_2$O$_3$ | 1.08 | 0.49 | 1.71 | 0.75 | 0.92 | 1.27 | 1.39 | 0.81 |
| La$_2$O$_3$ | 0.00 | 0.00 | 0.00 | 0.03 | 0.00 | 0.00 | 0.00 | 0.00 |
| Ce$_2$O$_3$ | 0.00 | 0.00 | 0.02 | 0.01 | 0.05 | 0.08 | 0.02 | 0.00 |
| Pr$_2$O$_3$ | 0.03 | 0.00 | 0.00 | 0.02 | 0.05 | 0.00 | 0.00 | 0.00 |
| Nd$_2$O$_3$ | 0.00 | 0.02 | 0.00 | 0.00 | 0.00 | 0.05 | 0.00 | 0.00 |
| Sm$_2$O$_3$ | 0.00 | 0.03 | 0.00 | 0.00 | 0.00 | 0.04 | 0.03 | 0.00 |
| Gd$_2$O$_3$ | 0.04 | 0.02 | 0.02 | 0.01 | 0.05 | 0.05 | 0.10 | 0.00 |
| Dy$_2$O$_3$ | 0.13 | 0.07 | 0.20 | 0.04 | 0.19 | 0.19 | 0.29 | 0.09 |
| Er$_2$O$_3$ | 0.20 | 0.05 | 0.20 | 0.16 | 0.09 | 0.14 | 0.26 | 0.00 |
| Yb$_2$O$_3$ | 0.37 | 0.11 | 0.33 | 0.21 | 0.35 | 0.42 | 0.56 | 0.31 |
| UO$_2$ | 0.57 | 0.47 | 1.25 | 0.47 | 1.80 | 2.14 | 0.40 | 0.54 |
| ThO$_2$ | 0.03 | 0.00 | 0.13 | 0.01 | 0.11 | 0.16 | 0.26 | 0.13 |
| PbO | 0.00 | 0.00 | 0.07 | 0.02 | 1.07 | 0.00 | 0.04 | 0.00 |
| F | 0.00 | 0.00 | 0.00 | 0.00 | 0.26 | 1.54 | 0.31 | 0.18 |

| | 1307-44 | 1310-14 | 735-5 | 1487-3 | 1283-22 | 1441-26 | 633-6 | 634-37 |
|---|---|---|---|---|---|---|---|---|
| O=F | 0.00 | 0.00 | 0.00 | 0.00 | 0.11 | 0.65 | 0.13 | 0.08 |
| Total | 100.22 | 100.05 | 100.54 | 99.99 | 95.31 | 93.55 | 92.97 | 98.84 |
| apfu (O=4) | | | | | | | | |
| Si | 0.974 | 0.980 | 0.974 | 0.981 | 0.957 | 0.775 | 0.949 | 0.970 |
| Al | 0.000 | 0.000 | 0.000 | 0.000 | 0.014 | 0.046 | 0.006 | 0.006 |
| Zr | 0.954 | 0.972 | 0.935 | 0.966 | 0.903 | 0.857 | 0.930 | 0.940 |
| Hf | 0.013 | 0.012 | 0.012 | 0.013 | 0.022 | 0.026 | 0.014 | 0.024 |
| Ca | 0.003 | 0.000 | 0.000 | 0.001 | 0.011 | 0.133 | 0.026 | 0.008 |
| Fe | 0.002 | 0.005 | 0.002 | 0.002 | 0.000 | 0.043 | 0.040 | 0.027 |
| Mn | 0.000 | 0.001 | 0.000 | 0.000 | 0.001 | 0.005 | 0.011 | 0.006 |
| Mg | 0.000 | 0.000 | 0.000 | 0.000 | 0.000 | 0.004 | 0.002 | 0.000 |
| P | 0.023 | 0.015 | 0.025 | 0.014 | 0.042 | 0.110 | 0.012 | 0.017 |
| Sc | 0.005 | 0.005 | 0.011 | 0.005 | 0.014 | 0.018 | 0.001 | 0.000 |
| As | 0.001 | 0.000 | 0.000 | 0.000 | 0.001 | 0.006 | 0.001 | 0.000 |
| Bi | 0.000 | 0.000 | 0.000 | 0.001 | 0.001 | 0.000 | 0.000 | 0.000 |
| Nb | 0.000 | 0.000 | 0.000 | 0.000 | 0.000 | 0.003 | 0.000 | 0.001 |
| W | 0.000 | 0.000 | 0.000 | 0.000 | 0.000 | 0.006 | 0.000 | 0.001 |
| Y | 0.018 | 0.008 | 0.028 | 0.012 | 0.016 | 0.023 | 0.025 | 0.014 |
| Dy | 0.001 | 0.001 | 0.002 | 0.000 | 0.002 | 0.002 | 0.003 | 0.001 |
| | 1307-44 | 1310-14 | 735-5 | 1487-3 | 1283-22 | 1441-26 | 633-6 | 634-37 |
| Er | 0.002 | 0.000 | 0.002 | 0.002 | 0.001 | 0.002 | 0.003 | 0.000 |
| Yb | 0.003 | 0.001 | 0.003 | 0.002 | 0.004 | 0.004 | 0.006 | 0.003 |
| U | 0.004 | 0.003 | 0.009 | 0.003 | 0.013 | 0.016 | 0.013 | 0.004 |
| Th | 0.000 | 0.000 | 0.001 | 0.000 | 0.001 | 0.001 | 0.005 | 0.001 |
| Pb | 0.000 | 0.000 | 0.001 | 0.000 | 0.009 | 0.000 | 0.000 | 0.000 |
| F | 0.000 | 0.000 | 0.000 | 0.000 | 0.054 | 0.336 | 0.066 | 0.036 |

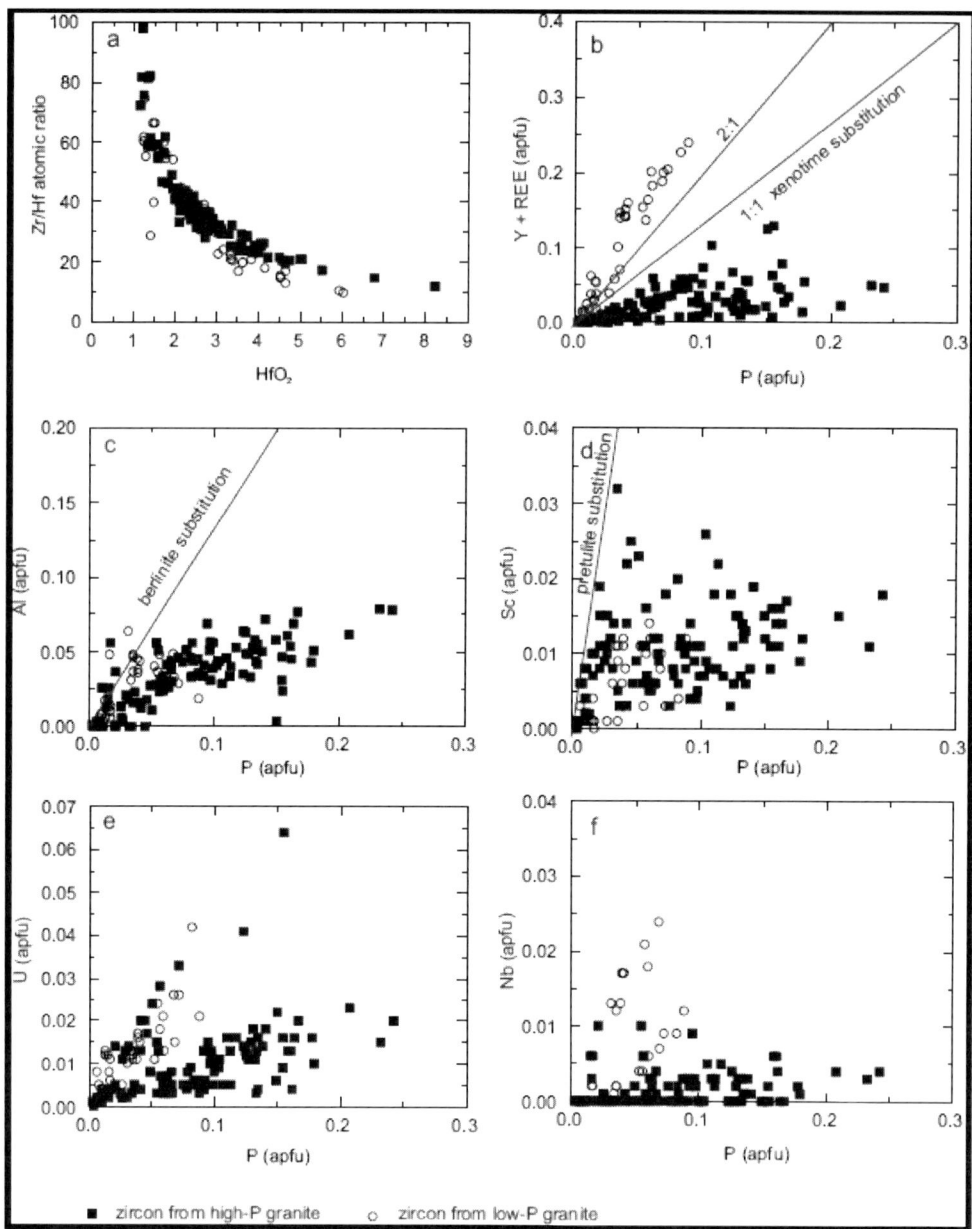

Figure 15. a) Plot of Zr/Hf vs. HfO$_2$ for zircon from P-poor and P-rich granites of the Krušné Hory/Erzgebirge area, b) Plot Y + REE vs. P for zircon from P-poor and P-rich granites of the Krušné Hory Erzgebirge area, c) Plot Al vs. P for zircon from P-poor and P-rich granites of the Krušné Hory/Erzgebirge area, d) Plot Sc vs. P for zircon from P-poor and P-rich granites of the Krušné Hory/Erzgebirge area, e) Plot U vs. Th for zircon from P-poor and P-rich granites of the Krušné Hory/Erzgebirge area, f) Plot Nb vs. P (apfu) for zircon from P-poor and P-rich granites of the Krušné Hory/Erzgebirge area.

In zircons from both topaz granite suites Th and U concentrations remain low and they are frequently below the detections limits. The highest observed Th concentration is 3.1 wt.% ThO$_2$ (0.03 apfu Th) in zircon from P-poor granites of the Cínovec cupola. However, the highest recognized U concentration was found in zircon from P-rich granite of the Horní

Slavkov–Krásno ore district (up to 7.7 wt.% $UO_2$; 0.06 apfu U). The U concentrations in zircons from P-poor granites of the Cínovec cupola are somewhat lower (4.9 wt.% $UO_2$; 0.04 apfu U) (Figure 15 e).

Altered zircons from topaz granites are also often enriched in Bi, As, Nb and W. The distribution of these four elements in zircon from investigated topaz granites is heterogeneous at every scale of observation. Zircons enriched in these elements occur together with zircons containing Bi, As, Nb and W below the microprobe detection limit. The highest Bi concentration (5.2 wt.% $Bi_2O_3$; 0.03 apfu Bi) was measured in zircons from partly greisenized P-rich granite occurring in the upper part of the Hub stock. The Bi concentrations in zircons from P-poor granites of the Cínovec cupola are distinctly lower (0–0.9 wt.% $Bi_2O_3$; 0–0.006 apfu Bi). The highest As concentrations were also found in partly greisenized P-rich granites of the Hub stock (up to 1.9 wt.% $As_2O_5$; 0.04 apfu As). The As concentrations in studied P-poor granites are partly lower (up to 1.2 wt.% $As_2O_5$; 0.02 apfu As). The concentrations of Nb and W are distinctly higher in P-poor granites (0.1–1.4 wt.% $Nb_2O_5$; 0–0.02 apfu Nb and 0–2.9 wt.% $WO_3$; 0–0.03 apfu W). Concentrations of both elements in P-rich granites from the Horní Slavkov-Krásno ore district are accordingly lower (0–0.6 wt.% $Nb_2O_5$; 0–0.009 apfu Nb and 0–1.3 wt.% $WO_3$; 0–0.01 apfu W) (Figure 15 f).

## CONCLUSION

The one coherent and cogenetic, ca. 400 km long plutonic megastructure of the Saxo-Danubian granite belt is formed by Fichtelgebirge/Erzgebirge compositional batholith in the Saxothuringian Zone and the South Bohemian batholith (SBB) in the Moldanubian Zone of the Central European Hercynian fold belt. Two-mica granites form an important part of the SBB. Two, mineralogically and geochemically contrasting granites, the Eisgarn and the Deštná granites, represent the majority of these granites. In the Eisgarn granite accessory minerals are concentrated in biotite flakes, whereas in the Deštná granite accessory minerals are predominantly enclosed in K-feldspar phenocrysts. These differences together with different Zr, Th and LREE concentrations in both granite types indicate a quite different melting history of both granites melts. Thorium, U, Y, Zr and REE-bearing accessory minerals in both two-mica granite types are represented by apatite, monazite, and zircon. In addition, in the Deštná granite rather rare xenotime-(Y) occurs. For xenotime-(Y) intimate intergrowths with zircon are significant. Zircons in both granite types contain a relatively low amount of $HfO_2$ (0.01–0.02 Hf apfu) and zircon from the Deštná granite is partly enriched in Y (up to 0.03 apfu). Monazite in both granite types is Ce-dominant (1.3–1.8 apfu Ce) with a restricted amount of cheralite component. In monazites from the Eisgarn granite the amount of cheralite components vary from 7.3 to 23.3 mol.%, whereas in monazites from the Deštná granite its content is distinctly lower, 8.8–12.6 mol.%. However, monazite from the Deštná granite is partly enriched in Y (up to 8.6 mol.% $YPO_4$). Xenotime-(Y) from the Deštná granite has a composition along the solid solution between xenotime and cheralite with predominance of U over Th (Th/U = 0.08–0.33). More dominant thorite exchange in xenotime from the Deštná granite is nearly absent due to distinctly low Th concentrations in this rock.

Topaz granites form two distinct magmatic suites in the Krušné Hory/Erzgebirge area. The first suite is formed by P-rich, highly evolved S-type granites. The second magmatic suite is formed by P-poor, A-type granites. Zircon, Nb-Ta-Ti oxides and uraninite are common accessory minerals in both topaz granite types. P-rich granites host large amounts of apatite, together with monazite and relative scarce xenotime. Apatite is absent in P-poor granites, and the low contents of whole-rock phosphorus are reflected by variable amounts of monazite and xenotime and by low phosphorus content in zircon. However, the distinctly high phosphorus content in P-rich granites are also reflected in feldspar, topaz and zircon (up to 8.3 wt.% $P_2O_5$). Both granite suites are characterized by zircon–xenotime intergrowths. Zircon grains very often show strong hydration and fluorination. All examined zircons contain moderate Hf concentration (1.2–8.2 wt.% $HfO_2$) with no considerable differences between both topaz granite suites. However, zircons from P-rich and P-poor granites show significant differences in the contents of Y. Zircons from the Cínovec granite cupola contain up to 18.4 wt.% $Y_2O_3$, whereas Y concentrations in zircon from analyzed P-rich granites reach only 5.5 wt.% $Y_2O_3$. Xenotime also occurs in both granite suites, but highly intimate zircon–xenotime intergrowths are present in P-poor granites from the Cínovec cupola. Epitaxial overgrowths of xenotime over zircon were frequently observed in these granites. Xenotime from the P-poor granite of the Cínovec cupola displays a considerable enrichment in Th (up to 4.2 wt.% $ThO_2$) and HREE (up to 35.7 wt.% $HREE_2O_3$) compared to xenotime from P-rich granites of the Horní Slavkov–Krásno ore district. The concentrations of Th in xenotimes from P-rich granites of the Horní Slavkov–Krásno ore district reach only 1.9 wt.% $ThO_2$. The concentrations of HREE in these granites are also distinctly lower, only up to 19.5 wt.% $HREE_2O_3$. However, analysed xenotimes from P-rich granites are distinctly enriched in U, up to 5.6 wt.% $UO_2$. Differences in Y, HREE, U and Th concentrations could also be displayed by substitution mechanisms occurring in studied xenotimes. For xenotime from P-rich granites the brabantite-type substitution is particularly significant, whereas xenotime from P-poor granites has a mostly thorite-type substitution.

## Acknowledgments

The analytical works of this study was supported by the Ministry of Education, Youth and Sports (project No. ME10083), the Czech Science Foundation (project No. 205/09/0540) and by institute research plan (AV0Z30460519) of Institute of Rock Structure and Mechanics of the CAS CR, public research institution. I am grateful to R. Škoda, Š. Benedová and P. Gadas from Institute of Geological Sciences of the Masaryk University (Brno) for technical assistance by microprobe analysis of selected minerals.

## References

[1] Wark, D. A.; Miller, C. F. *Chem. Geol.* 1993, 110, 49–67.
[2] Casillas, R.; Nagy, G.; Pantó, G.; Brändle, J.; Fórizs, I. *Eur. J. Mineral.* 1995, 7, 989–1006.
[3] Bea, F. *J. Petrol.* 1996, 37, 521–552.

[4] Förster, H-J. (1998a) *Amer. Mineral.* 1998, 83, 259–272.
[5] Förster, H-J. (1998b) *Amer. Mineral.* 1998, 83, 1302–1315.
[6] Hoskin, P.W.O.; Kinny, P. D.; Wyborn, D.; Chappell, B. W. *J. Petrol.* 2000, 41, 1365–1396.
[7] Johan, Z.; Johan, V. *Mineral. Petrol.* 2005, 83, 113–150.
[8] Franke, W. In Orogenic Processes: Quantification and Modelling in the Variscan Belt; Franke, W.; Haak, V.; Oncken, O.; Tanner, D.; Ed.; Geol. Soc. London Spec. Publ. 179; *Geol. Soc.* London: London, 2000; pp 35–61.
[9] Friedl, G.; Finger, F.; McNaughton, N.; Fletcher, *I. R. Geology* 2000, 28, 1035–1038.
[10] Winchester, J.A.; *PACE TMR Network Team Tectonophysics* 2002, 360, 5–21.
[11] Finger, F. Gerdes, A. René, M. Riegler, G. *Geol. Carpath.* 2009, 60, 205–212.
[12] Förster, H-J.; Romer, R. L. In Pre-Mesozoic Geology of Saxo-Thuringia. Linnemann, U.; Romer, R. L.; Ed.; *Schweizerbart Science Publishers*: Stuttgart, 2010; pp 297–308.
[13] Siebel, W.; Chen, F.; Satir, M. *Int. J. Earth Sci.* 2003, 92, 36–53.
[14] Siebel, W.; Shang, C. K.; Reitter, E.; Rohrmüller, J. Breiter, K. *J. Petrol.* 2008, 49, 1853–1872.
[15] Gerdes, A.; Wörner, G.; Henk, A. *J. Geol. Soc.* London 2000, 157, 577–587.
[16] Finger, F.; Clemens, J. Contrib. Mineral. Petrol. 1995, 120, 311–326.
[17] Verner, K.; Žák, J.; Pertoldová, J.; Šrámek; J.; Sedlák, J.; Trubač, J.; Týcová, P. *Int. J. Earth Sci.* 2009, 98, 517–532.
[18] Vellmer, C.; Wedepohl, K. H. Contrib. *Mineral. Petrol.* 1994, 118,13–32.
[19] Holub, F.; Klečka, M.; Matějka, D. In Pre-Permian geology of Central and Eastern Europe; Dallmeyer, R. D.; Franke, W.; Weber, K.; Ed.; *Springer Verlag*: Berlin, 1995; pp 444–452.
[20] Breiter, K; Gnojek, I.; Chlupáčová, M. Věst. *Ústř. Úst. Geol.* 1998, 73, 301–312.
[21] René, M.; Holtz, F.; Luo, Ch.; Beermann, O.; Stelling, *J. Lithos* 2008, 102, 538–553.
[22] Breiter, K.; Förster, H-J.; Seltmann, R. Mineral. Deposita 1999, 34, 505–521.
[23] Förster, H-J.; Tischendorf, G.; Trumbull, R. B.; Gottesmann, B. *J. Petrol.* 1999, 40, 1613–1645.
[24] Breiter, K. Lithos 2011, doi: 10.1016/j.lithos.2011.09.022.
[25] Müller, A.; René, M.; Behr, H-J.; Kronz, *A. Mineral. Petrol.* 2003, 79, 167–191.
[26] Taylor, R. P.; Fallick, A. E. *Terra Nova* 1997, 9, 105–108.
[27] René, M.; Škoda, R. Mineral. *Petrol.* 2011, 103, 37–48.
[28] Seifert, T.; Kempe, U. Beih. *Eur. J. Miner.* 1994, 6, 125–172.
[29] Štemprok, M.; Šulcek, Z. *Econ. Geol.* 1969, 64, 392–404.
[30] Dolejš, D.; Štemprok, M. Bull. Czech *Geol. Surv.* 2001, 76, 77–99.
[31] Pouchou, J. L.; Pichoir, F. In Microbeam analysis. Armstrong, J. T.; Ed.; San Francisco Press: San Francisco, 1985; pp 104–106.
[32] Chappel, B. W.; Hine, R. *Res. Geol.* 2006, 56, 203-244.
[33] Sylvester, P.*J. Lithos* 1998, 45, 29-44.
[34] Taylor, S. R.; McLennan, S. M. The continental crust: its composition and evolution. Blackwell: Oxford, 1985; pp 1–312.
[35] Maniar, P. D.; Piccoli, P. M. *Geol. Soc. Amer. Bull.* 1989, 101, 635–643.

[36] Whalen, J. B.; Currie, K. L.; Chappell, B. W. Contrib. Mineral. *Petrol.* 1987, 95, 407–419.
[37] Pabst, A.;Hutton, C. O. *Amer. Mineral.* 1951, 36, 60–69.
[38] Bowie, S. H. U.; Horne, J. E. T. *Mineral. Mag.* 1953, 30, 93–99.
[39] Linthout, K. *Can. Mineral.* 2007, 45, 503–508.
[40] Friedrich, M. H.; Cuney, M. In Uranium deposits in magmatic and metamorphic rocks; *IAEA*: Vienna, 1989; pp 11–35.
[41] Ward, C. D.; McArthur, J. M.; Walsh, J. N. J. Petrol. 1992, 33, 785–815.
[42] Montel, J-M. *Chem. Geol.* 1993, 110, 127–146.
[43] Harlov, D.E.; Procházka, V.; Förster, H-J.; Matějka, D. *Mineral. Petrol.* 2008, 94, 9–26.
[44] Hetherington, C. J.; Jercinovic, M. J.; Williams, M. L.; Mahan, K. *Chem. Geol.* 2008, 254, 133–147.
[45] Wang, R. C.; Fontan, F.; Chen, X. M.; Hu, H.; Liu, C. S.; Xu, S. J.;. Parseval, P. de. *Can Mineral*, 2003, 41, 727–748.
[46] Wang, R. C.; Wang, D. Z.; Zhao, G. T.; Lu, J.J.; Chen, X. M.; Xu, S. J. Phys. *Chem. Earth* 2001, A26, 835–849.
[47] Förster, H-*J. Lithos* 2006, 88, 35–55.
[48] Uher, P.; Breiter, K.; Klecka, M.; Pivec, E. *Geol. Carpath.* 1998, 49, 151–160.
[49] Wang, R. C.; Zhao, G. T.; Lu, J. J.; Chen, X. M.; Xu, S. J.; Wang, Z. *Mineral. Mag.* 2000, 64, 867–877.
[50] Kempe, U.; Gruner, T.; Renno, A.; Wolf, D.; René, M. *Mineral. Mag.* 2004, 68, 669–675.
[51] Breiter, K.; Förster, H-J.; Škoda, R. *Lithos* 2006, 88, 15–34.
[52] Pupin, J. P.Trans. Roy. Soc. Edinburgh Earth. Sci. 2000, 91, 245–256.
[53] Hoskin, P. W. O.; Schaltegger, U. The composition of zircon and igneous and metamorphic petrogenesis. In Zircon; Hanchar, J. M.; Hoskin, P. W. O.; Ed.; Reviews in Mineralogy and Geochemistry 53; *Mineral. Soc. Amer.*: Wahington, 2003; pp 27–62.
[54] Soba-Pérez, C.; Villaseca, C.; González del Tánago, J.; Nasdala, L. *Can. Mineral.* 2007, 45, 509–527.
[55] Hoshiho, M.; Kimata, M.; Nishida, N.; Shimizu, M.; Akasaka, T. N. Jb. *Miner. Abh.* 2010, 187, 167–188.
[56] Whitney, D. L.; Evans, B. W. *Amer. Mineral.* 2010, 95, 185–187.

In: Granite
Editors: Miroslava Blasik and Bogdashka Hanika
ISBN: 978-1-62081-566-3
© 2012 Nova Science Publishers, Inc.

**Chapter 3**

# CRUST AND MANTLE CONTRIBUTIONS TO OROGENIC GRANITOID MAGMATISM: THE CASE OF THE TERTIARY MAGMATISM OF ALPS (ITALY)

*Laura Pinarelli*[1,*], *Angelo Peccerillo*[2,†], *and Carmelita Donati*[2]

[1]Istituto di Geoscienze e Georisorse, Sezione di Firenze,
Via G. La Pira, Firenze, Italy
[2]Dipartimento di Scienze della Terra, Università degli Studi di Perugia,
Piazza dell'Università, Perugia, Italy

## ABSTRACT

Tertiary orogenic magmatic rocks occur extensively along the Alps. Exposed lithologies are dominated by intermediate to acid granitoids but also include some dikes and volcanics. Major, trace element, and isotopic signatures show important variations along the Alps, highlighting complex origin and differentiation processes. Calcalkaline to shoshonitic magmatism is ubiquitous, whereas ultrapotassic alkaline rocks are restricted to the Western Alps. A few peraluminous granites, representing pure crustal anatectic magmas, are associated to some of the major intrusive bodies.

Individual plutons, such as Bregaglia and Adamello, exhibit positive covariation of whole rock $^{87}Sr/^{86}Sr$ vs. $\delta^{18}O$, a feature indicative of magma-wall rock interaction. However, some incompatible element ratios (e.g., Th/Nb, La/Nb, Th/Y) do not change much during this process, suggesting that these ratios may be used to constrain compositions of parental magmas. Mantle normalized incompatible element patterns of mafic rocks (MgO >3 wt%) are fractionated and enriched in Large Ion Lithophile Element (LILE: Rb, K, LREE, U, Th, Pb) and relatively depleted in High Field-Strength

---

[*] E-mail: lapina@igg.cnr.it.
[†] E-mail: pecceang@unipg.it.

Elements (HFSE: Ta, Nb, Zr, Hf, Ti). The lowest LILE abundances and LILE/HFSE ratios are observed in the mafic rocks from Adamello.

Sr-Nd isotope ratios are negatively correlated both at the regional scale and within single granitoid bodies. The highest Sr- and lowest Nd-isotope ratios are observed in the Western Alps, where some mafic potassic alkaline dykes show crustal-like isotopic signatures. Pb-isotope ratios have more restricted ranges than Sr-Nd isotopes, and mostly fall in the compositional field of the upper crust. Low Sr- and high Nd-isotope ratios and relatively low Pb-isotope signatures are observed in some Adamello mafic rocks in the Central Alps, reflecting the contribution from relatively uncontaminated mantle sources.

Overall, geochemical and isotopic evidence suggests that both crust and mantle end-members contributed to the origin and evolution of Alpine Tertiary magmatism. Some compositional features for individual plutons, such as covariation of Sr and O isotopic ratios, are related to local processes of magma contamination by crustal wall rocks. In contrast, other features such as regional distribution of petrological types and variation of LILE/HFSE and Sr-Nd-Pb isotopes of mafic rocks along the Alps, are better explained by heterogeneous mantle sources that had been variably contaminated by fluids released from subducted upper crustal material during Africa-Europe convergence. Occurrence in the Western Alps of ultrapotassic rocks with very high Sr- and low Nd-isotope ratios suggests a stronger degree of mantle metasomatism by upper crust in this sector. By contrast, lower LILE/HFSE ratios and lower Pb-isotope ratios at Adamello highlight stronger role of pristine mantle end-members, which were not contaminated by subducted crust. Such a disparity of crustal contribution to magmatism is related both to variation of local stress regimes and to modification of attitude and nature of the subducting slab during Africa-Europe collision.

## 1. INTRODUCTION

Oligocene to Miocene magmatic activity is a main geological feature along the Alps. This was a consequence of convergence and collision of the Europe and Africa plates, with southward subduction of the former occurring beneath the North African continental margin (e.g., Dal Piaz & Venturelli, 1983; Alagna et al., 2010 with references; Dal Piaz, 2010 with references; Handy et al., 2010). Magmatism took place both during subduction, and during phases of post-collisional extensional tectonic regime (e.g., Dal Piaz & Venturelli, 1983; Beltrando et al., 2011 and references therein). Extrusive rocks were extensively eroded and are mostly found as detrital andesitic clasts within sedimentary deposits (e.g., Ruffini et al., 1997), and as a few remnants of lava flows and pyroclastic rocks (e.g., Altherr et al., 1995; Callegari et al., 2004). Most of the Oligo-Miocene magmatism is, therefore, intrusive and its study provides important insight into processes of magma generation and evolution at converging plate margins, and on the origin of orogenic granitoids.

Previous studies (e.g., Ulmer et al., 1983; Bart et al., 1989; Kagami et al., 1991; Bellieni et al., 1996; von Blanckenburg et al., 1998; Alagna et al., 2010 with references) have pointed out that intrusive magmatism in the Alps exhibits strong petrological and geochemical variations, with important modifications of rock composition both at the scale of individual intrusive bodies and at the regional scale along the chain. There is a wide consensus that the ultimate origin of the Alpine magmatism is a mantle wedge modified by subduction processes (e.g., Ulmer et al., 1983; Kagami et al., 1991). Additionally, there is much compelling petrological and geochemical evidence that magmas underwent strong compositional modification during emplacement into the crust. Therefore, petrological and geochemical complexities of Alpine magmatism has been attributed to the cumulative effects of regional compositional modifications of the primary mantle-derived magmas as well as different types and degrees of magma evolution processes (i.e., fractional crystallisation, combined assimilation plus fractional crystallization - AFC, and magma mixing) within the crust during emplacement. However, the relative role played by each these factors is still to be defined. Such an issue is common to most if not all orogenic granitoid provinces at global scale.

In this paper we review the main compositional characteristics of the Tertiary intrusive magmatism along the Alps, point out the regional geochemical variations of the magmas, discuss possible effects of intra-crustal processes for key suites where a sufficient number of data is available, explore implications for primary magma compositions at the regional scale, and discuss contributions of mantle and crust to magma genesis along the Alpine converging continental margin.

## 2. GEOLOGICAL OUTLINE

The Cainozoic magmatism along the Alps (here termed "Alpine magmatism") is a product of the convergence between the Africa and Europe plates and the consequent subduction of the European lithosphere beneath Adriatic block, a promontory of the northern African crustal margin. This process of subduction consumption of the Tethys oceanic plate has been occurring since Cretaceous-Eocene time, with continental collision commencing during the Oligocene (e.g., Rosenbaum and Lister, 2005; Handy et al., 2010).

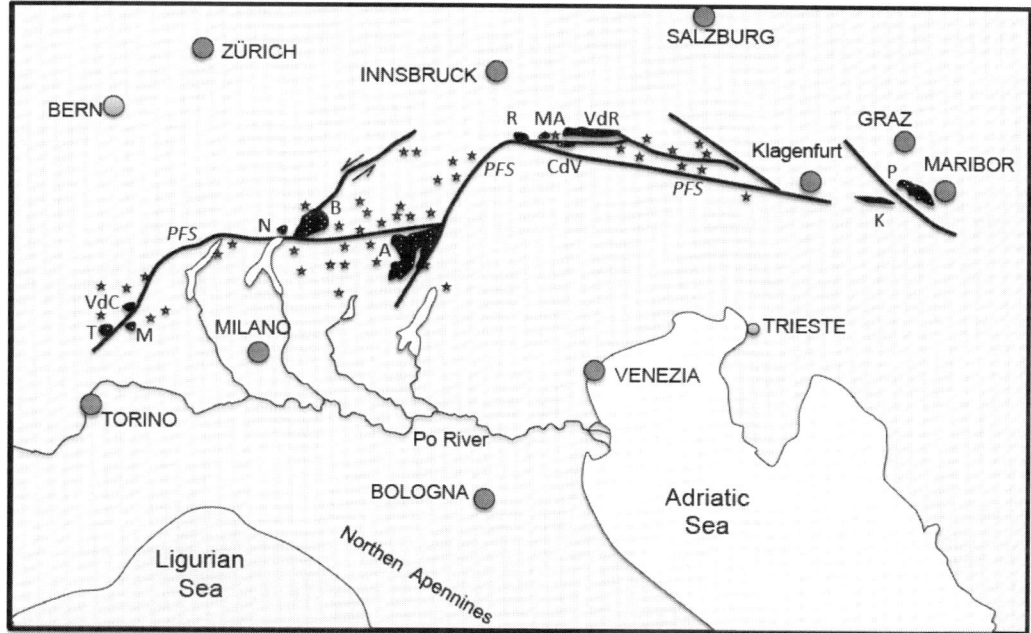

Figure 1. Schematic location map of Alpine intrusions. T = Traversella; VdC = Valle del Cervo; M = Miagliano; N = Novate; B = Bregaglia (Bergell); A = Adamello; R = Rensen; MA = Monte Alto; VdR = Vedrette di Ries; CdV = Cima di Vila; K = Karavanke; P = Pohorje. Stars indicate locations of some dikes and minor intrusive bodies. PFS = Periadriatic Fault System.

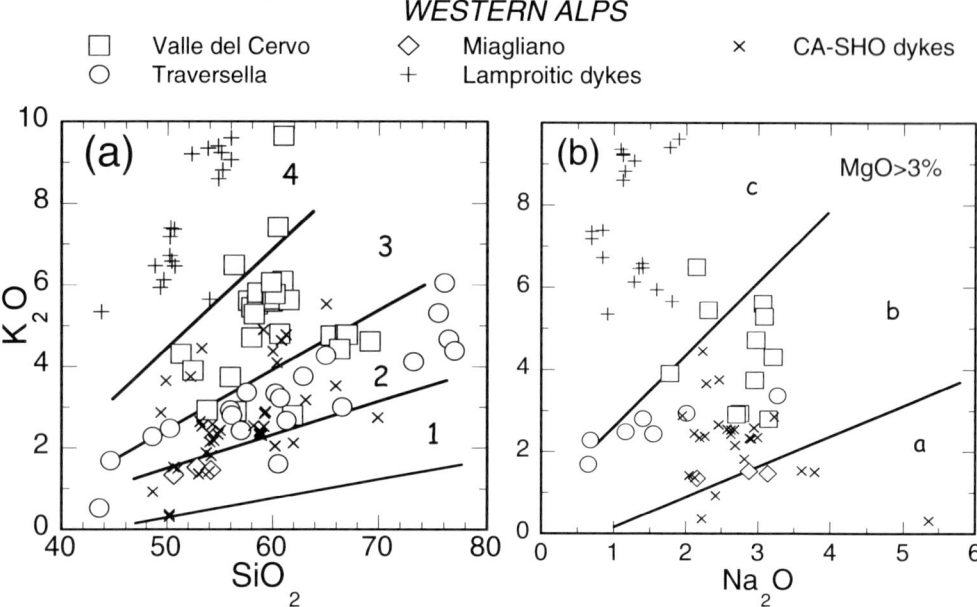

Figure 2. (a) $K_2O$ vs. $SiO_2$ and (b) $K_2O$ vs. $Na_2O$ classification diagrams for the Tertiary plutonic bodies of the Western Alps. Ultrapotassic lamproitic dikes from the same area are also noted. Lines in Figure 2a divide fields of calcalkaline (1), high-K calcalkaline (2), shoshonitic (3), and ultrapotassic (4) series. Lines in Figure 2b divide mafic rocks (MgO > 3wt%) with sodic (a), potassic (b), and ultrapotassic (c) affinities.

Alpine magmatism occurred between about 43 to 24 Ma, with a climax around ~ 32-30 Ma (e.g., Berger and Bousquet, 2008; Stampfli & Hochard, 2009; Handy et al., 2010). Granitoid intrusions were mainly emplaced along the Periadriatic Fault System (Figure 1), a large transpressive tectonic lineament that extends from the Western Alps up to the Pannonian Basin (e.g., Rosenberg, 2004; Dal Piaz, 2010). In addition to plutonic bodies, a large number of dykes were emplaced. These represent either minor satellite intrusions of the main plutonic bodies, or independent batches of magmas.

Plutonic bodies mainly consist of tonalites and granodiorites plus subordinate amounts of gabbros, diorites, and granites, all of which have a calcalkaline (CA) to high-K calcalkaline (HKCA) petrochemical affinity. Minor amounts of potassic alkaline rocks are also present. Dikes are mainly calcalkaline andesites, but shoshonites and ultrapotassic lamproites also occur. The latter are restricted to the Western Alps (Venturelli et al., 1984; Peccerillo and Martinotti, 2006; Conticelli et al., 2009).

## 3. PETROLOGICAL AND GEOCHEMICAL CHARACTERISTICS OF THE MAIN PLUTONIC BODIES

The main Oligo-Miocene plutonic bodies present along the Alps (Figure 1) include the Traversella and Valle del Cervo intrusions in the Western Alps, the Bregaglia (Bergel) and Adamello plutons in the Central Alps, and the Rensen, Monte Alto (Altenberg), Vedrette di Ries (Rieserferner), and Cima di Vila (Zinsnock) in the Eastern Alps. The Karavanke and

Pohorje plutons represent the easternmost occurrence of the Alpine intrusions in Austria and Slovenia.

Traversella has an age of about 30 Ma and consists of predominant diorite and minor monzonite, with several granitic and aplitic dykes crossing the main plutonic body (van Marcke de Lummen & Vander Auwera, 1990). The petrologic affinity of rocks is HKCA to shoshonitic (Figure 2), with the most mafic rocks (MgO >3 wt%) exhibiting $K_2O/Na_2O$ of 1 or higher. Mantle-normalised incompatible element patterns of these mafic rocks (not shown) are positively fractionated with high concentrations of LILE (Rb, Ba, Th, U, LREE), and relative depletion in Ti, Zr and HREE (see Alagna et al., 2010).

Initial Sr-Nd-Pb isotopic ratios of diorites and monzonites (Figures 3 and 4) have restricted ranges ($^{87}Sr/^{86}Sr \sim 0.7099$-$0.7107$; $^{143}Nd/^{144}Nd \sim 0.5122$, $^{206}Pb/^{204}Pb \sim 18.67$-$18.69$; $^{207}Pb/^{204}Pb \sim 15.64$-$15.67$; $^{208}Pb/^{204}Pb \sim 38.73$-$38.86$). Oxygen isotope compositions of whole-rocks and some separated minerals increase with $SiO_2$ from the near-mantle values of $\delta^{18}O = +5.9‰$ for some mafic cumulates, to $\delta^{18}O = +10.5‰$ for some granitic dykes.

Major, trace element and isotope variations suggest that the Traversella pluton was generated by a process of assimilation plus fractional crystallization (AFC), starting from mantle-derived parental magmas. Mafic cumulates record the most primitive isotopic compositions of the pluton, and indicate an origin by melting of anomalous mantle sources that were contaminated by subducted upper crustal material (van Marcke de Lummen & Vander Auwera, 1990).

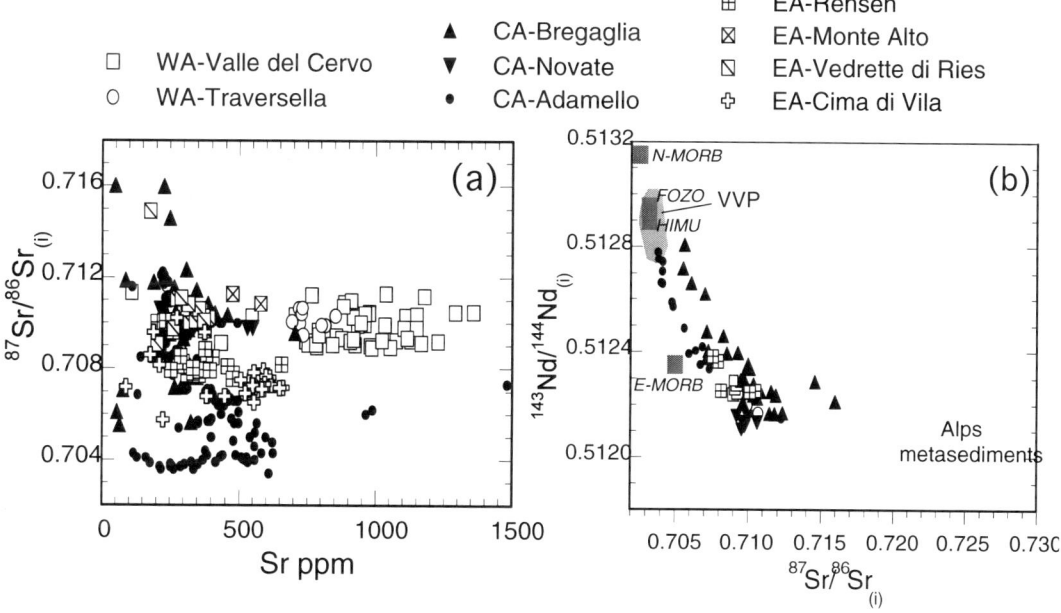

Figure 3. Diagrams of (a) $^{87}Sr/^{86}Sr$ vs. Sr and (b) $^{143}Nd/^{144}Nd$ vs. $^{87}Sr/^{86}Sr$ for Alpine Tertiary plutonic rocks. Compositional fields of metasediments from Western Alps (Voshage et al., 1987; 1990, Pinarelli et al., 2008) and for the Veneto Volcanic Province (VVP) (Peccerillo & Martinotti, 2006; Lustrino & Wilson, 2007; Macera et al. 2008) are also shown in (b). The metasediments from Alps represent possible contaminants of both magma and mantle, whereas the VVP represent an OIB-type source in the area. MORB, HIMU, and FOZO after Stracke et al. (2005).

The Valle del Cervo pluton (~31-29 Ma) has a monzogranite to granodiorite complex at the core, with syenitic-monzonitic rocks in the external zones. Several small satellite bodies with varied compositions (monzogabbro, gabbronorite, quartz-diorite, monzodiorite, syenite) occur around the main intrusion. The Miagliano stock (monzonite to diorite) is considered the largest of these bodies.

The rocks of the Valle del Cervo pluton comprise a shoshonitic association, but some monzonitic and syenitic rocks have a nearly ultrapotassic affinity with $K_2O/Na_2O >2$ (Figure 2b). Incompatible element patterns of mafic rocks are variably fractionated with enrichments in LILE (e.g., Cs, Th, U, Pb) and strong relative depletions of Nb, Ta and Ti. Both LILE and HFSE (e.g., Rb, Ba, Th, U, Nb, Sr, Zr, LREE) increase with $K_2O/Na_2O$ for rocks with the same degree of compositional evolution. Overall, LILE enrichments at Valle del Cervo are significantly higher than at Traversella.

Initial Sr-isotope ratios of the Valle del Cervo intrusion ($^{87}Sr/^{86}Sr \sim 0.7089$-$0.7113$) are consistently high (Figure 3a). Lower values are observed in the Miagliano stock ($^{87}Sr/^{86}Sr \sim 0.7068$-$0.7075$; Carraro and Ferrara, 1968). Initial Nd-isotope ratios (Figure 3b) have a restricted range ($^{143}Nd/^{144}Nd \sim 0.51225$ and $0.51228$). Ratios of $^{207}Pb/^{204}Pb$ (15.61-15.70), $^{208}Pb/^{204}Pb$ (38.60-38.87) and $206Pb/204Pb$ (18.72-18.79) fall in a narrow compositional range (Figure 4).

Major, trace element and isotopic data show that rocks at Valle del Cervo have more variable compositions than at Traversella. Various magmas evolved from a suite of isotopically and geochemically diverse parental melts by AFC, and intruded as distinct magma batches in different zones of the pluton, from the centre to the periphery. The parental magmas were generated by melting of a heterogeneous upper mantle source, which had been contaminated by subducted upper crustal material (Bigioggero et al., 1994). The most potassic rocks have affinities with some lamproitic dykes occurring in the Western Alps, but the enrichments in incompatible elements are lower at Valle del Cervo.

The Bregaglia (Bergell) pluton, with an age of ~33–30 Ma, consists predominantly of tonalites and granodiorites plus minor amounts of gabbro and mafic to felsic dykes. Granite is the dominant rock of the Novate stock, which is located at the western margin of the Bregaglia pluton. Rocks are mostly calcalkaline and high-K calcalkaline in composition (Figure 5a), with some mafic dykes extending to tholeiitic, potassic, and ultrapotassic alkaline affinities (Figure 5b). Incompatible element patterns (normalized to primordial mantle composition) of mafic rocks are enriched in LILE and relatively depleted in HFSE. Potassic alkaline dikes are more enriched in incompatible elements than CA and HKCA rocks, and imply a derivation from a distinct type of parental magma.

Initial Sr- and Nd-isotope ratios of the Bregaglia pluton are very variable ($^{87}Sr/^{86}Sr = 0.7055$-$0.7160$; $^{143}Nd/^{144}Nd = 0.51225$-$0.51272$) and exhibit the usual negative correlation (Figure 3b). $^{206}Pb/^{204}Pb$ ratios (18.73-18.83) are similar to those of Valle del Cervo pluton, whereas $^{207}Pb/^{204}Pb$ and $^{208}Pb/^{204}Pb$ (15.67-15.74 and 38.75-39.00, respectively) are slightly higher.

Major, trace element and isotopic data of the Bregaglia rocks have been interpreted as evidence that parental magmas were formed by partial melting of an enriched lithospheric mantle that had been contaminated by subducted crustal material. Compositional evolution processes within the crust were dominated by AFC (von Blanckenburg et al., 1992). Tiepolo et al. (2002) proposed a magma origin by melting of spinel-bearing MORB-type mantle, metasomatized during Alpine and/or pre-Alpine times. Shoshonitic lamprophyres were formed at deeper levels in the mantle, within the field of garnet-lherzolite. The Novate granite is considered as a distinct body with respect to Bregalia, derived by partial melting within the crust (von Blanckenburg et al., 1992; Liati et al., 2000). Mingling textures between leucogranite and diorite-tonalite indicates that crustal melting and the generation of felsic magmas in the Novate granite are related to the intrusion and heating of the crust by mantle-derived magmas of the Bregaglia body.

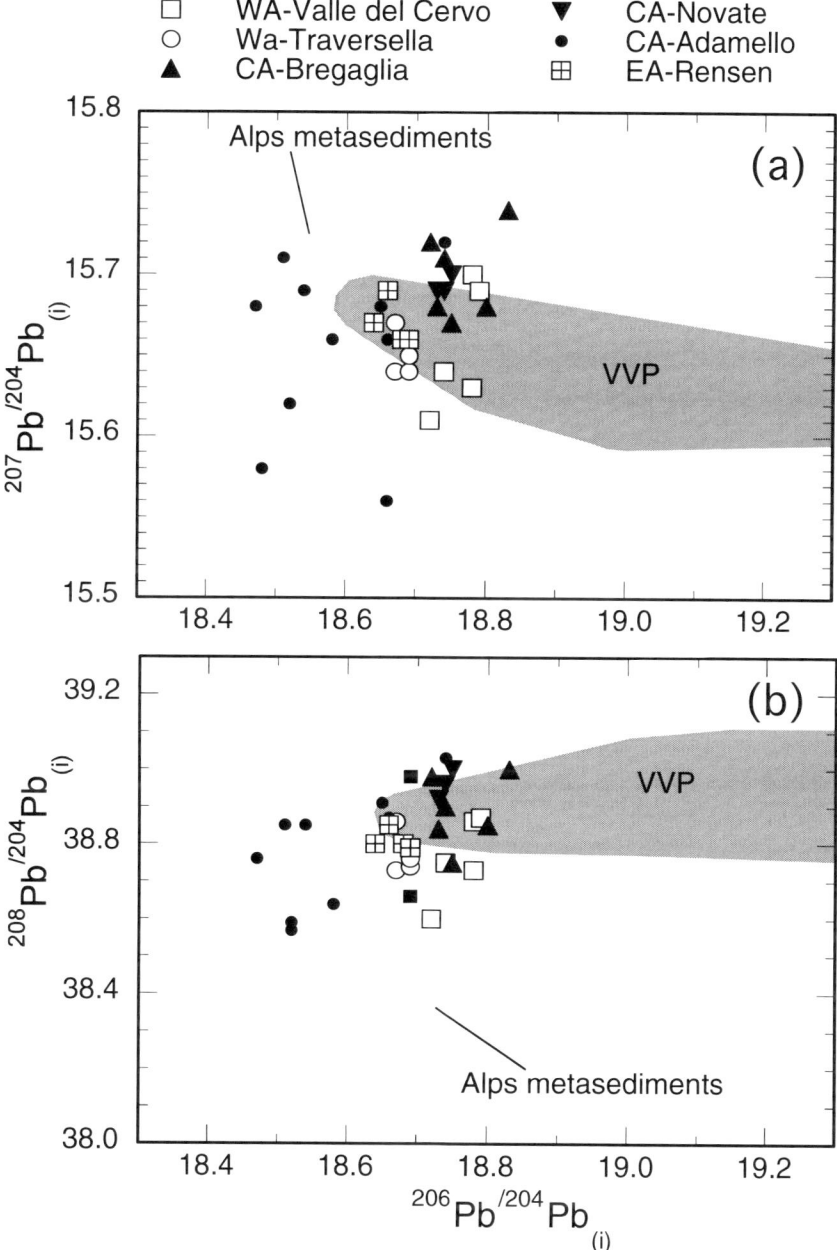

Figure 4. Diagrams of $^{207}Pb/^{204}Pb$ (a) and $^{208}Pb/^{204}Pb$ (b) vs. $^{206}Pb/^{204}Pb$ showing the Alpine Tertiary plutonic rocks. Also shown are the fields of metasediments from Western Alps (Cumming et al., 1987; Pinarelli et al., 2008) and of VVP (Macera et al., 2008; Lustrino & Wilson, 2007).

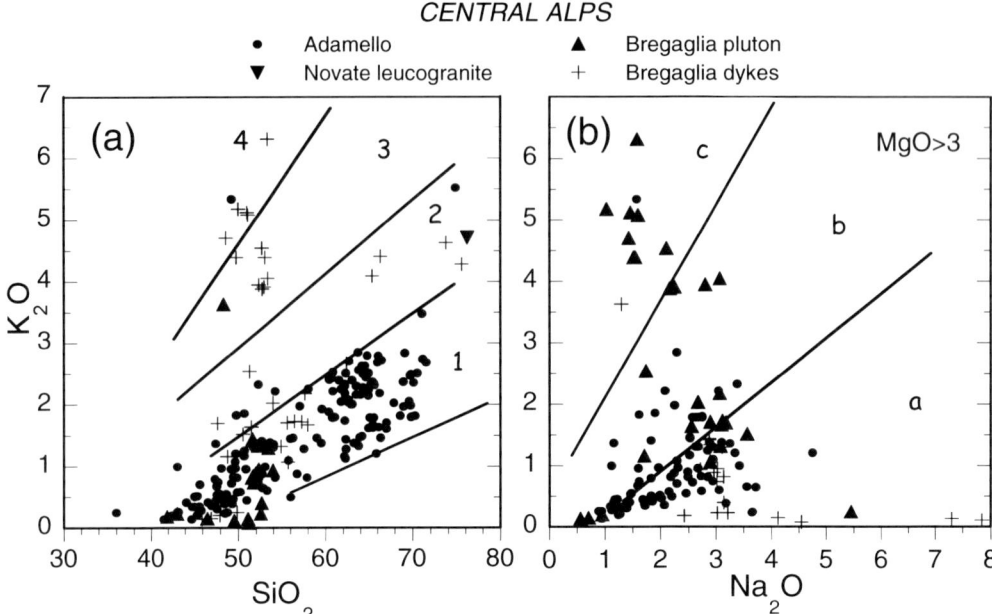

Figure 5. Classification diagrams of (a) $K_2O$ vs. $SiO_2$ and (b) $K_2O$ vs. $Na_2O$ for mafic rocks for the Tertiary plutons of the Central Alps. Lines in Figure 5a divide fields of calcalkaline (1), high-K calcalkaline (2), shoshonitic (3), and ultrapotassic (4) series. Lines in Figure 5b divide mafic rocks with sodic (a), potassic (b), and ultrapotassic (c) affinities.

The Adamello batholith is the largest and best studied of the Alpine Tertiary intrusions. It is a composite pluton that consists of several coalesced bodies: Re di Castello, Adamello, Avio, and Presanella (Callegari, 1983). The ages of intrusion are in the range of 43 to 29 Ma, and decrease from south to north. The rocks consist of dominant tonalite and granodiorite, plus gabbro, diorite, quartz-diorite, trondhjemite and granite. Some ultramafic rocks of cumulate origin (wherlites, hornblendites) are also present. These were emplaced at the intersection between two main tectonic structure, the Giudicarie and the Periadriatic lines (Handy et al., 2010), which caused local extensional tectonics and favoured magma intrusion.

Chemical compositions of Adamello rocks are calcalkaline to high-K calcalkaline, with potassium increasing from the southern to the northern plutons (Figure 5a). Some mafic rocks have a tholeiitic affinity. Mafic rocks (MgO >3%) exhibit variably fractionated mantle-normalised incompatible trace elements patterns with relative enrichments in Ba, Rb, Th, U, Pb and depletion in Nb and Ti.

Sr-isotope ratios display a tendency to increase with K and Rb at the scale of the batholith. Therefore, they increase from south to north in parallel to K and Rb. The southern intrusion of Re di Castello has the lowest Sr isotope ratios ($^{87}Sr/^{86}Sr \sim 0.703$-$0.707$), whereas the northern Adamello, Avio and Presanella plutons ($^{87}Sr/^{86}Sr \sim 0.707$-$0.712$) have higher values (Figure 3a). Cortecci et al. (1979) reported a positive correlation of $^{87}Sr/^{86}Sr$ with oxygen isotopes ($\delta^{18}O‰ \sim +5.9$ to $+9.4‰$) for whole rocks. $^{143}Nd/^{144}Nd$ ratios (0.51278 to 0.5121, Figure 3b) are negatively correlated with Sr isotopes. Pb-isotope ratios (Figure 4) have a larger range of values than other Alpine intrusions, and some samples exhibit relatively less radiogenic $^{206}Pb/^{204}Pb$ and $^{207}Pb/^{204}Pb$ ratios of 18.48-18.74 and 15.56-15.72, respectively (Dupuy et al., 1982; Kagami et al., 1991).

The Adamello composite pluton had a complex origin and evolution. Positive $^{87}Sr/^{86}Sr$ vs. $SiO_2$ correlations observed for some intrusions, has been interpreted as evidence for AFC-dominated processes for single magma bodies. This agrees with the positive correlation of Sr and O isotopes (Cortecci et al., 1979). The northward increase of K, Rb and Sr-O isotope ratios has been attributed to an overall increasing role of AFC from south to north (Cortecci et al., 1979; Dupuy et al., 1982; Del Moro et al., 1983a, b; Kagami et al., 1991). Parental magmas have an Mg-tholeiitic to picritic composition, similar to some mafic dikes (Ulmer et al., 1983), and were generated in a garnet-lherzolitic upper mantle, contaminated by fluids released from subducted oceanic crust (Ulmer et al., 1983). Tiepolo and Tribuzio (2005) suggested a source contamination by subducted sediments.

The Rensen pluton has an age of 31.7 to 30.8 Ma (Barth et al., 1989). It consists mainly of granodiorites with minor tonalite and quartz-diorite (Bellieni et al., 1984; Barth et al., 1989; Bellieni et al., 1991) that mostly display a calcalkaline affinity (Figure 6a). Primordial mantle-normalized trace element patterns of mafic rocks are, as other intrusions, fractionated with moderate enrichments of LILE and LREE, and negative anomalies of Nb and Ti.

Sr-isotope ratios (Figure 3a) in the mafic-intermediate rocks exhibit moderate variability ($^{87}Sr/^{86}Sr \sim 0.7075$-$0.7081$), whereas some evolved felsic bodies have much more radiogenic compositions ($^{87}Sr/^{86}Sr \sim 0.7095$-$0.7110$). Nd-isotope ratios (0.51236-0.51225) show the usual negative correlation with Sr isotopes (Figure 3b). Pb-isotope ratios fall near the narrow field defined by the Traversella plutonic rocks (Figure 4).

Like the other Alpine igneous rocks, the Rensen plutonic suite has been interpreted as the result of extensive interaction between mantle and continental crust (Barth et al., 1989; Bellieni et al., 1991). Bellieni et al. (1991) argued for a complex multistage polybaric fractional crystallisation.

The Monte Alto pluton is located east of the Rensen. Formed at around 24-25 Ma (Borsi et al., 1979; Sassi et al., 1985), it consists predominantly of granodiorite plus some tonalite (Bellieni et al., 1984). The Monte Alto stock has HKCA petrochemical affinity (Figure 6a) and has higher enrichments in LILE than the equivalent rocks from Rensen. Initial $^{87}Sr/^{86}Sr$ ratios ($\sim 0.7108$-$0.7113$, Figure 3a) are similar to slightly higher than those of the Rensen felsic rocks. Geochemical and isotope data indicate that Monte Alto magmas originated by melting of plagioclase-free, garnet-bearing deep crustal rocks, with some fractionation during emplacement (Bellieni et al., 1984).

The Vedrette di Ries pluton formed around 32 Ma old and consists mainly of granodiorite and tonalite, with small lenses and dykes of diorite and granite. Most rocks are garnet-bearing (Bellieni et al., 1981, 2010 with references) and display a HKCA affinity (Figure 6a). Mantle-normalised incompatible trace element patterns of mafic rocks show high enrichments in LILE and the usual negative anomalies of Nb, Ta, Ti and Sr.

Sr-isotope ratios display a moderate range of variation ($^{87}Sr/^{86}Sr \sim 0.709$-$0.711$, Figure 3a), with only some aplitic dikes reaching high $^{87}Sr/^{86}Sr$ ratios of $\sim 0.715$. Trace element and Sr-isotope data indicate that the Vedrette di Ries magmas derived from mantle-generated parental melts that underwent a two-stage fractional crystallization process, at different pressures, both accompanied by crustal assimilation (Bellieni et al., 1981).

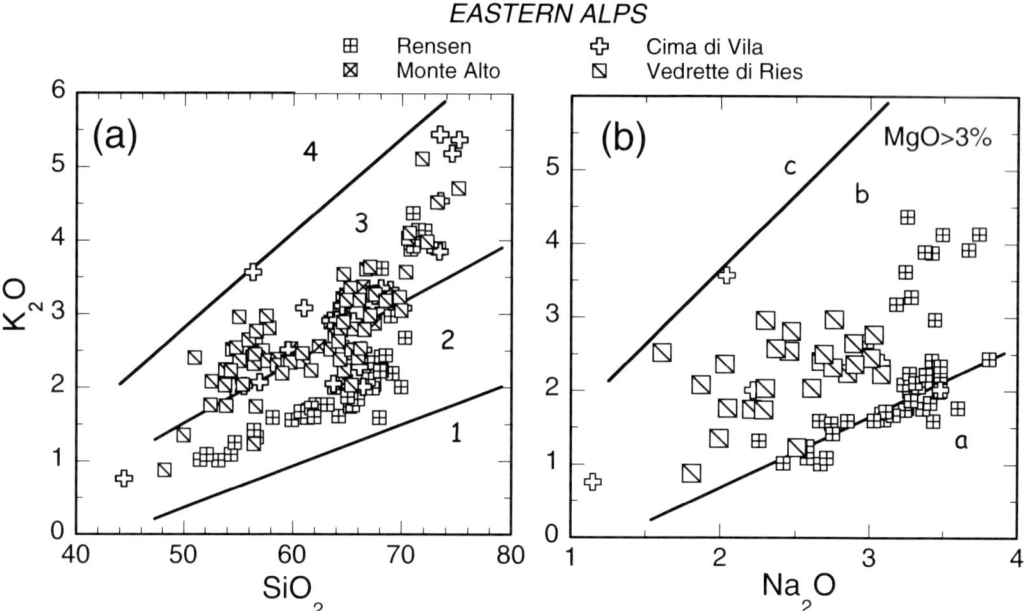

Figure 6. Classification diagrams for $K_2O$ vs. $SiO_2$ and (b) $K_2O$ vs. $Na_2O$ for mafic rocks for Eastern Alps Tertiary plutonic intrusions. Lines in Figure 6a divide fields of calcalkaline (1) and high-K calcalkaline (2) series. Lines in Figure 6b divide mafic rocks with sodic (a), potassic (b), and ultrapotassic (c) affinities.

The Cima di Vila (Zinsnock) intrusion was emplaced at about 29.5 Ma, close to the southern border of the Vedrette di Ries pluton. It is formed predominantly of granodiorite and some tonalite, with granitic and aplitic dikes crosscutting the magmatic body (Bellieni et al., 1996, 2010). Cima di Vila rocks display a HKCA petrochemical affinity (Figure 6a) and their major element chemistry is very similar to that of the Vedrette di Ries pluton. However, the range of Sr-isotope ratios at Cima di Vila is much wider and overall lower ($^{87}Sr/^{86}Sr$ ~ 0.7058-0.7102, Figure 3a) than at Vedrette di Ries (Bellieni et al., 1996). According to Bellieni et al. (1996) the Cima di Vila magmatism is the result of mixing between mafic mantle-derived and felsic crustal melts, the latter being generated within the lower crust. However, the most acid rocks represent products of fractional crystallization starting from hybrid magmas.

The Karavanke and Pohorje plutons (28-30 Ma), which crop out in Austria and Slovenia, are the easternmost intrusions of the Periadriatic magmatic system. These poorly studied intrusions consist mainly of tonalites plus minor granodiorites and diorites that have a CA and HKCA affinity (Pamic & Palinkas, 2000, with references therein). Their major element composition is similar to that of equivalent rocks from other Alpine intrusions. Initial Sr-isotope ratios are moderately high at both Pohorje, $^{87}Sr/^{86}Sr$ ~ 7066-0.7075, and Karavanke, $^{87}Sr/^{86}Sr$ ~ 0.7075 (Pamic & Palinkas, 2000). Geochemical and isotopic data suggest that these plutons were formed by AFC from tholeiitic basalt parental magmas that were generated in a slightly metasomatized garnet peridotite (Pamic & Palinkas, 2000; Kovacs et al., 2007).

## 4. DISCUSSION

Geochemical data show that interaction between mantle and crustal end-members is a first-order process in the origin and chemical evolution of the Alpine magmatism. This is clearly demonstrated by variations in radiogenic isotope ratios (Figures 3 and 4), which are intermediate between crust and mantle. Most authors agree that mantle-crust interaction occurred both by introduction of upper crustal material into the upper mantle via subduction (source contamination) and during emplacement of magmas into the continental crust (magma contamination). The two processes generate similar effects on the isotopic composition of magmas, and compositions observed in the Sr-Nd-Pb isotope space are the cumulative effects of both magma and source contamination.

### 4.1. Intra Crustal Magma Evolution and Source Contamination

Understanding modifications undergone by magmas during evolutionary processes within the crust is critical to better constrain the compositions of primary melts and the characteristics of their mantle sources. Such an objective can be reached only by performing detailed geochemical and isotopic studies on individual plutonic bodies through the analysis of both whole rocks and separated phases. The scarceness of these studies for the Alpine magmatism makes it difficult to discuss the nature and effects of intra-crustal magma assimilation processes, and to provide firm constraints on compositions of parental melts.

Combined Sr-O isotopic investigations are generally considered as the best tool to discriminate between magma and source contamination and to obtain reliable indications of the amounts of crustal material involved (e.g., Harmon et al., 1984; Harmon and Hoefs, 1995). However, this type of data is available for a limited number of samples from a few Tertiary Alpine intrusions. A diagram of Sr- and O-isotope variations for the Bregaglia and Adamello plutons from the data of Diethelm (1990), von Blanckenburg et al. (1992), and Cortecci et al. (1979) is shown in Figure 7. Covariations of Sr and O define positive correlations for both rock suites. In principle, such behaviour could result either from mantle contamination by subducted upper crustal material or from magma contamination during emplacement. However, mantle contamination is able to modify very significantly Sr isotopic compositions, whereas $\delta^{18}O$ is less affected (dashed line in Figure 7). In contrast, magma contamination produces a stronger increase of the O than Sr isotopes (solid lines).

Most of the data reported in Figure 7, plot along a simple two end-member mixing line between a calcalkaline basaltic composition (e.g., Adamello or Bregaglia) and a typical upper continental crust. AFC models run somewhat at lower O isotope values, with some samples from Bregaglia plotting on AFC1 trend characterised by a high $m_a/m_c$ (mass of assimilated over mass of crystallised material) value of about 0.5. Overall, O-Sr isotope evidence points to significant upper crustal involvement in magma evolution, as already suggested by previous studies (e.g., Cortecci et al., 1979; von Blanckenburg et al., 1992). Simple geochemical modelling suggests various styles of magma-crust interaction, from simple magma-crust mixing to AFC with rather high ratio of assimilated vs. crystallised material. However, it has to be stressed that such a conclusion is based on a limited number of data and has to be considered as broadly qualitative; only models based on detailed studies of single plutons and their wall rocks can provide reliable quantitative constraints on magma-wall rock interaction.

Some additional insight about magma–crust interaction can be gained by combined Sr-Nd isotope modelling, as shown in Figure 8. The most primitive (gabbroic) magmas from Adamello and Bregaglia have Sr- and Nd-isotope compositions quite close to typical mantle-derived melts. Bregaglia diorites have Nd-isotope ratios similar to those of the most primitive Adamello rocks, but are displaced towards slightly more radiogenic Sr-isotope compositions. These differences are difficult to model by magma contamination processes and, therefore, most likely reflect compositions of primary melts. Sr-Nd isotope data of both plutons define hyperbolic mixing trends that have mantle-derived calcalkaline magmas and metasediments from the Alps at their extremities. This again indicates strong magma contamination during emplacement. Sr-Nd isotope data for many samples plot along two-end-member mixing curves, as already observed for Sr-O isotope modelling.

Pb-isotope ratios for the various plutons define more restricted ranges of variation than do Sr-Nd isotopes, with most samples falling in the field of Alpine metasediments. Such decoupling of the Pb vs. Sr and Nd isotopes is a common process during mantle-crust interaction, since the strong enrichment in Pb of many upper crustal rocks, with respect to many mantle-derived melts, induces crustal Pb isotopic signatures in the hybrid magmas even for low degrees of crustal contamination (see Pinarelli et al., 1993 for discussion). Overall, Pb-isotope ratios of the Alpine plutons define vertical trends between mantle and crust in the $^{207}Pb/^{204}Pb$ vs. $^{206}Pb/^{204}Pb$ diagram (Figure 9), providing additional support to a role for both mantle and crust in magma origin and differentiation.

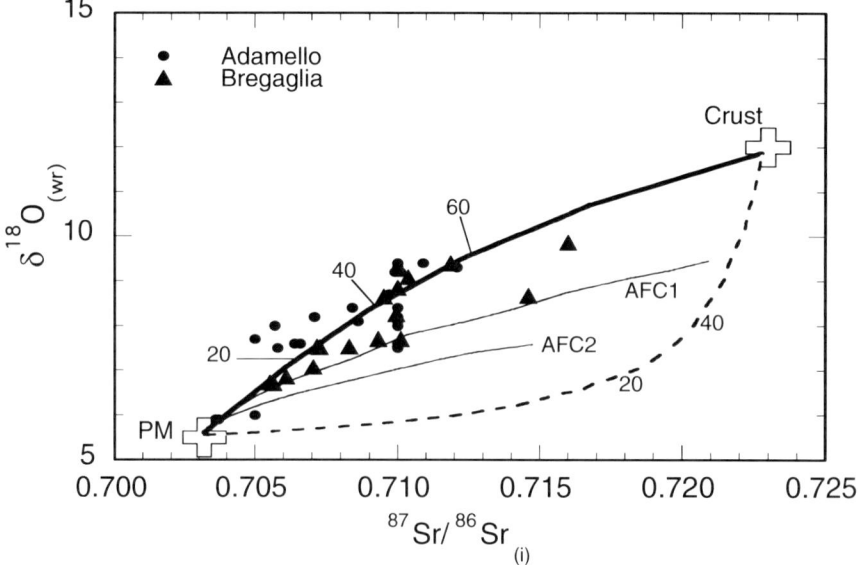

Figure 7. Variation diagram of $^{87}Sr/^{86}Sr$ vs. $\delta^{18}O$ for the Bregaglia and Adamello plutons. The dashed line shows the evolutionary path produced by mixing between primordial mantle and average metasediments from Alps. The heavy solid line denotes the trajectory followed by mixing between a mafic calcalkaline magma from Adamello and average metasediments from Alps. The thin solid lines show two different paths for combined assimilation-crystal fractionation between a mafic calcalkaline magma from Adamello and average metasediments from Alps ($D_{Sr} = 1.2$; $m_a/m_c = 0.5$ (AFC1) and 0.3 (AFC2)). Numbers along the lines represent the amount of added crustal material.

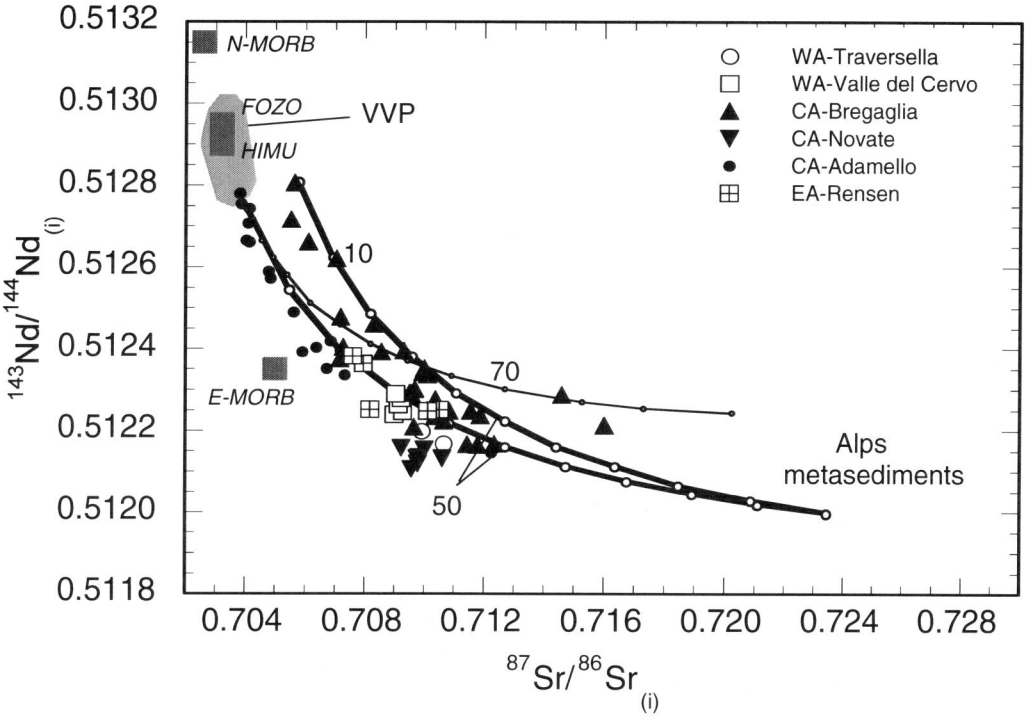

Figure 8. Diagram of $^{143}$Nd/$^{144}$Nd vs. $^{87}$Sr/$^{86}$Sr for Alpine Tertiary plutonic rocks showing the calculated models of mantle-crust interaction. Heavy solid lines illustrate the effect of mixing between mafic calcalkaline magmas from Adamello and Bregaglia, and average metasediments from Alps. The thin solid line shows trend for combined assimilation-fractional crystallization for a mafic calcalkaline magma from Adamello assimilating average metasediments from Alps ($D_{Sr}$ = 1.2; ma/mc = 0.5; $D_{Nd}$ = 0.1). Numbers along the lines represent the amount of added crustal material.

The samples with lower Pb-isotope ratios come from Adamello. Starting from these magmas, the increase of $^{207}$Pb/$^{204}$Pb in the Alpine rocks can be modelled by the addition of variable amounts of crustal materials. Both mixing and AFC processes can fit the distribution of the Alpine intrusions, inasmuch as either model produces almost identical trends in the Pb-Pb diagrams. Models show that the addition of only 10% metasediments is sufficient to shift the Pb-isotope ratios of a mafic magma into the field of crustal rocks. Amounts of added metasediments from about 10 to 40% is sufficient to explain the composition of most Alpine rocks, although some compositions with extreme $^{207}$Pb/$^{204}$Pb require the addition of very large quantities of crust (up to 70%). Note, however, that the assimilant can change significantly for various intrusions and the values reported above have to be considered as broadly qualitative. Initial magma compositions of modelling have somewhat different Pb-isotope signatures, which may reflect derivation from heterogeneous mantle sources.

In conclusion, radiogenic isotope compositions of Alpine Oligo-Miocene magmatism show strong evidence of mantle-crust interaction. Available Sr-O isotopic data suggest that much of the observed isotopic variations are the effect of intra-crustal magma contamination. The dominant processes seem to have been both bulk mixing between mantle- and crust-derived melts, and AFC with high contamination over crystallisation, a conclusion that requires validation by detailed studies on single rock suites.

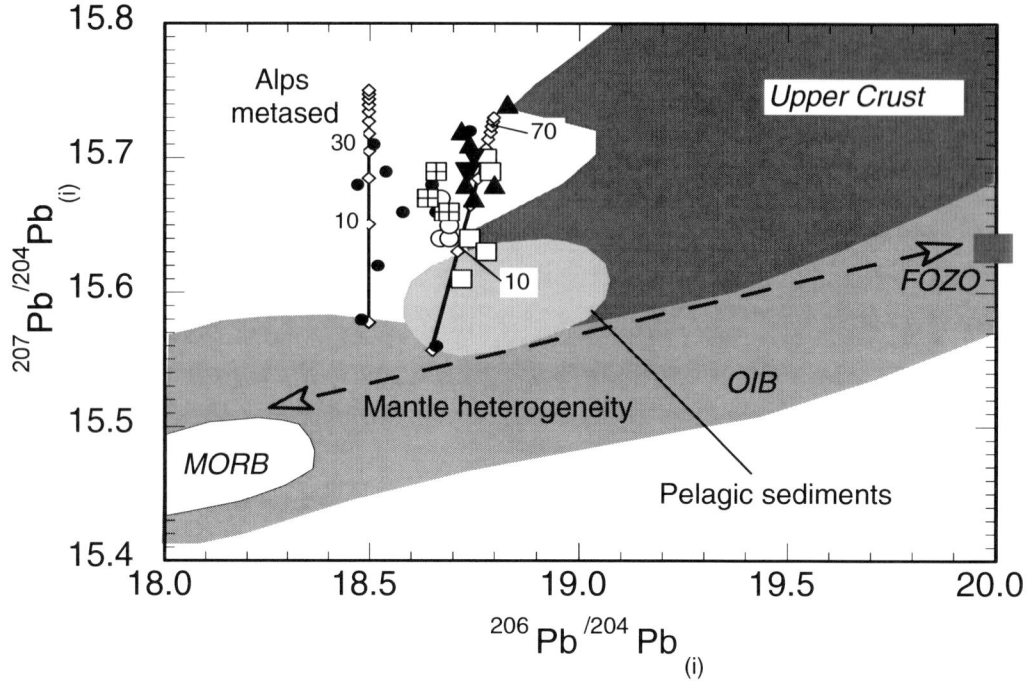

Figure 9. Diagram of $^{207}Pb/^{204}Pb$ vs. $^{206}Pb/^{204}Pb$ showing Tertiary Alpine plutons and the composition of metasediments from Western Alps (Cumming et al., 1987; Pinarelli et al., 2008). The fields of Upper Crust and Pelagic Sediments are after Zartman & Doe (1981); MORB, OIB, and FOZO after Zartman & Doe (1981) and Stracke et al. (2005). Solid lines are calculated mixing models between two quartz-diorite magmas from Adamello and two different metasediments from Strona-Ceneri. Numbers along the lines indicate the amount of added metasediment. The range of Pb-isotope variation for Alpine magmatism is best explained by heterogeneous mantle-derived calc-alkaline magmas that assimilated a heterogeneous metasedimentary crust.

The Sr- and Pb-isotope characteristics observed in the mafic rocks of some plutons (i.e., Pb-isotope signatures of Adamello and distinct Sr-isotope signatures of Adamello and Bregaglia) are difficult to be explained by magma contamination and may indicate the occurrence of distinct types of primary melts, which could reflect an origin in heterogeneous mantle sources.

A main problem of this study of Alpine magmatism is to establish the extent of trace element modification by magma-crust interaction. This is crucial for understanding which elements, if any, preserved compositions of parental melts and can be used to have information on mantle sources beneath the Alpine chain. Some indications on this issue can be obtained from diagrams that plot key trace element ratios vs. O isotope ratios of whole rocks, which is the best parameter to detect magma assimilation. Figure 10 shows the variation of some LILE/HFSE ratios vs. O isotope ratios for Traversella, Bregaglia and Adamello. Except for Rb/Nb of Adamello, trace element ratios do not show significant covariation with O isotopes. This suggests that the considered element ratios were not heavily affected by magma-wall rock interaction and can be used to detect regional variations of magmas. Therefore, their variation at the regional scale, especially for mafic rocks, may reflect the heterogeneity of mantle sources.

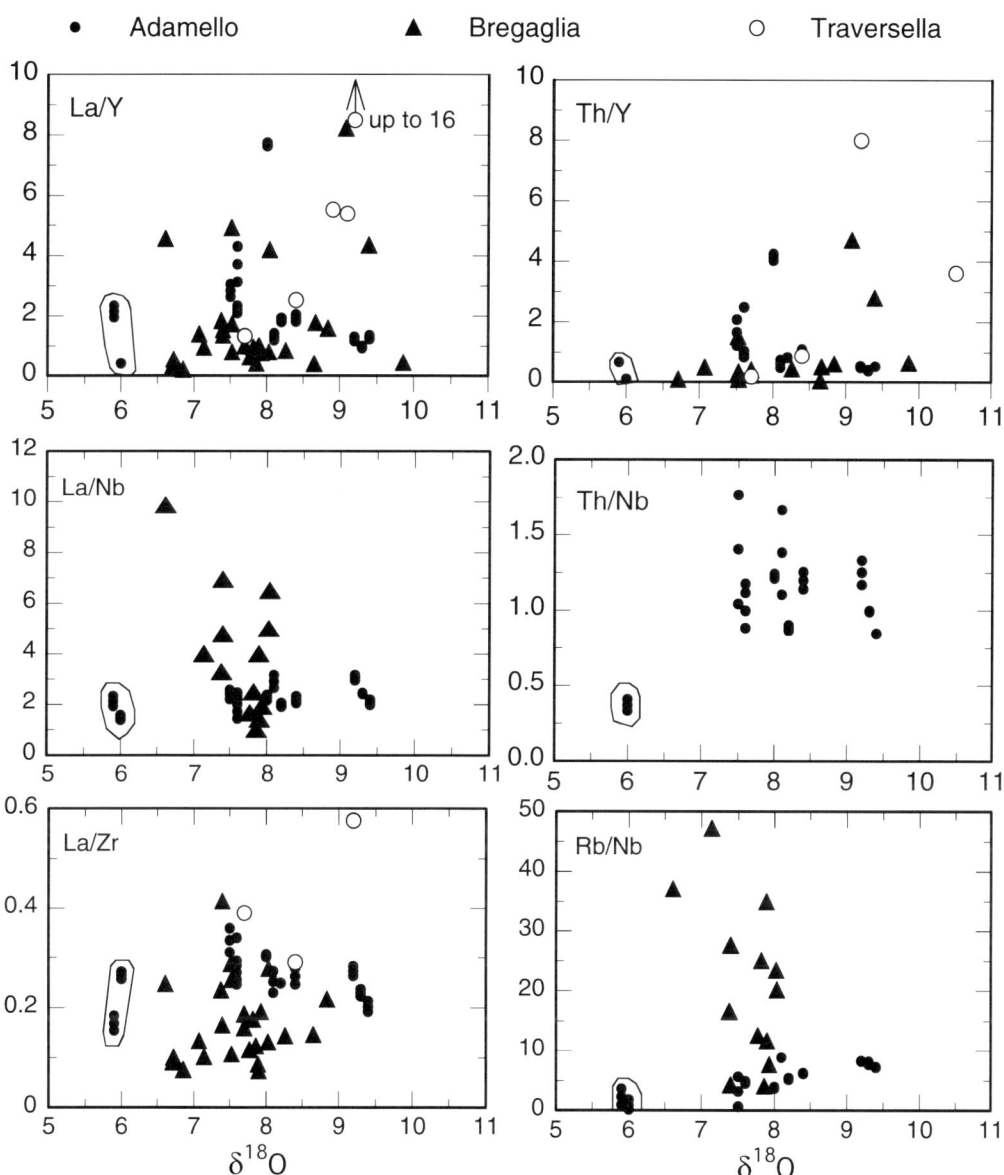

Figure 10. Plots of LILE/HFSE vs. $\delta^{18}O$ for Bregaglia and Adamello. Mafic samples (MgO > 3 wt%) of Adamello are circled. The samples from Traversella are also shown.

## 4.2. Regional Variations of Trace Element and Isotope Compositions

Plots of La/Nb vs. Th/Y and of Th/Nb vs. $^{87}Sr/^{86}Sr$ for the mafic rocks (MgO > 3%) of some Alpine plutons are shown in Figure 11. The compositional fields of lamproitic dikes from Western Alps, of metasedimentary rocks from Alps, of the Veneto Volcanic Province (VVP), and of MORB are also displayed. The VVP can be considered as a representative of an OIB-type source in the area (Macera et al., 2008). The metasediments from Alps represent possible contaminants for both magma and mantle.

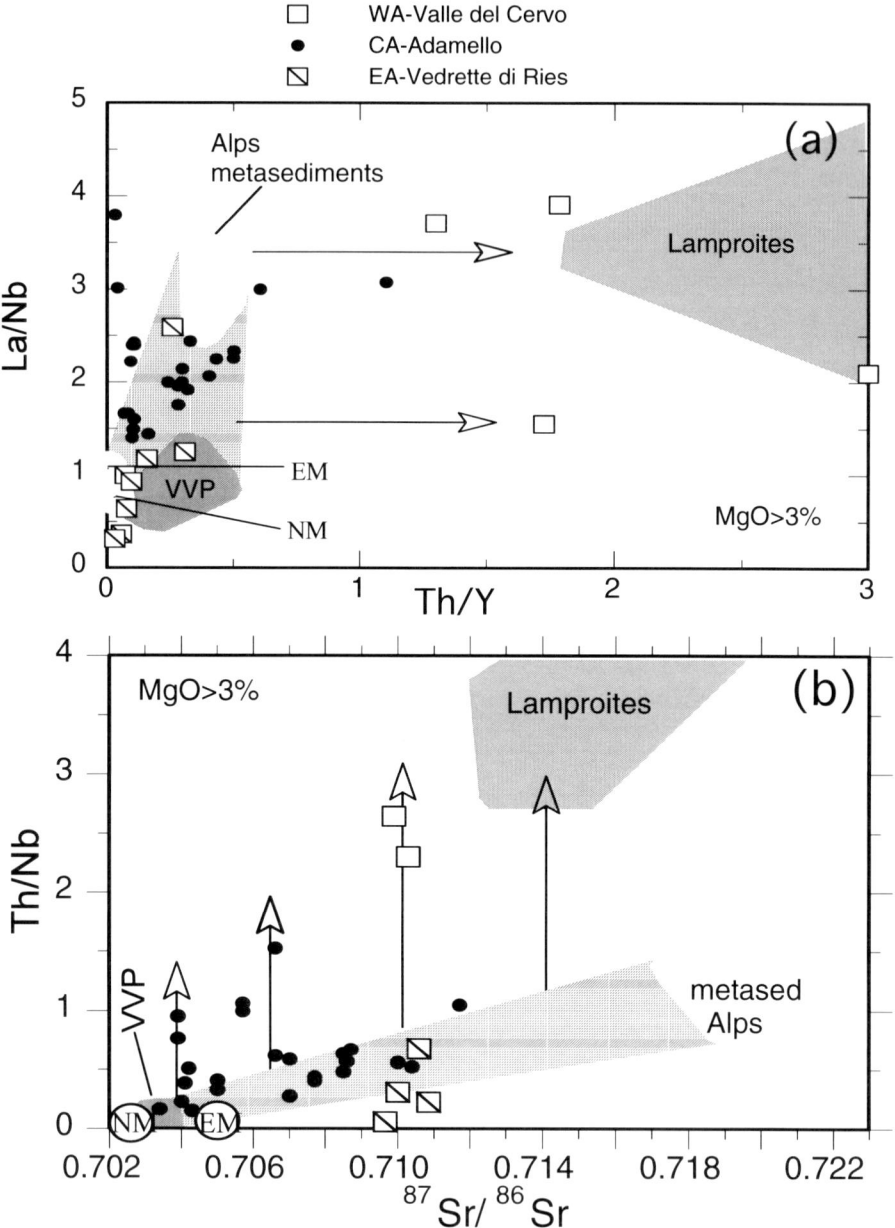

Figure 11. Diagrams of La/Nb vs. Th/Y (a) and Th/Nb vs. $^{87}Sr/^{86}Sr$ (b) for Alpine Oligo-Miocene intrusive rocks. The field of Alpine metasediments, Western Alps lamproitic dikes, and of the Eocene-Miocene Veneto Volcanic Province are also shown. N-MORB (NM) and E-MORB (EM) from Sun & McDonough (1989) and Stracke et al. (2005).

Most mafic Alpine plutonic rocks exhibit compositions that plot between metasediments from Alps and the upper mantle (VVP and MORB). This is indicated in Figure 11 as a ruled area, which broadly represents mixed compositions between mantle and crustal end-members. If the considered element ratios are pristine or poorly modified values of primary melts, as suggested earlier, the ruled area in Figure 11 should be considered as indicative of mantle contamination by subducted metasediments.

Adamello shows the largest compositional variation, with some samples showing VVP- or MORB-type values of both Sr isotopes and LILE/HFSE ratios, and others plotting in the mixed compositions between mantle and Alpine sediments. Mantle-like LILE/HFSE and $^{87}Sr/^{86}Sr$ ratios shown by some Adamello mafic rocks are unique for the entire Alpine magmatism, on the basis of available data. This indicates important contributions of uncontaminated mantle in the origin of this pluton. It should be noted that Adamello is the largest Tertiary intrusion in the Alps and contains notable amounts of mafic rocks. All these characteristics are probably related to the particular stress regime of the intrusion area, which favoured mantle melting at the intersection of Giudicarie and Periadriatic lithospheric fault systems.

Some samples of mafic plutonic rocks do not plot in the mantle-metasediment mixing area and are displaced toward higher LILE/HFSE ratios (Th/Y, Th/Nb in Figure 11). Very high Th/Nb and other LILE/HFSE ratios are also shown by Western Alps lamproites.

Lamproites are considered as the Alpine rocks whose mantle sources were most heavily contaminated by subducted upper crustal material (e.g., Peccerillo and Martinotti, 2006). This is strongly supported by their highly radiogenic Sr-isotope signatures, which are not far from those of metasediments. However, lamproites have much higher Th/Y and Th/Nb (and other LILE/HFSE) than metasediments (Figure 11a and b), indicating selective elemental enrichments during mantle metasomatism and/or melting. In particular, high Th/Y may be caused by the occurrence of residual garnet either in the slab during element transfer from sediments to mantle wedge, or during mantle melting that generated lamproitic magmas. Note that garnet is the only main rock-forming mineral that has much higher partition coefficients for Y and HREE than for LILE. Such an explanation, however, cannot be applied to high Th/Nb, which is not affected by residual garnet. High Th/Nb and other LILE/HFSE ratios of many rocks from various plutons can be attributed to the particular style of mantle enrichment during subduction. Several studies (e.g., Kessel et al., 2005) have shown that LILE are preferentially enriched with respect to HFSE during element transfer from slab to mantle wedge by aqueous fluids. Therefore, high Th/Nb values in many Alpine intrusive rocks, as well as in the Western Alps lamproites, may depend on such a process, highlighted by vertical arrows in Figure 11b, and visible for rocks of diverse compositions, independently on the degree of contamination of their mantle sources. Some rocks from Valle del Cervo, in the Western Alps, show very high Th/Y and Th/Nb, suggesting the occurrence of a lamproitic component in their source. This is supported by the consistently high Sr-isotope ratios of Valle del Cervo pluton (Figure 3a).

## CONCLUSION

Oligo-Miocene intrusive rocks of the Alps exhibit significant variations of major, trace element, and both oxygen and radiogenic isotope ratios. Overall, these document intensive interaction between mantle and crustal end-members in the origin and compositional evolution of these rock suites. A part of this interaction can be attributed to the introduction of upper crustal material into the mantle wedge during subduction of the European plate beneath the North African margin (source contamination). Another part is related to magma contamination during emplacement into the continental crust (magma contamination).

Available data are insufficient to the aim of unravelling the complex interplay of different factors in the origin of Alpine magmatism, and the conclusions reported in this paper should be considered as working hypotheses that need validation by further and much detailed studies.

Modelling of available isotopic data indicate that interaction of magma with wall occurred either by bulk assimilation of crustal material by intruding magmas or by AFC with high amounts of assimilation. These processes produced strong modifications of isotopic and geochemical data. However, variation of radiogenic isotope ratios observed for mafic rocks of some plutons cannot be derived from magma contamination but may instead reflect source heterogeneity. The low-Sr isotope ratios and LILE/HFSE in some Adamello rocks, with respect to other intrusions, suggest stronger role of uncontaminated mantle sources in the origin of this pluton. This can be attributed to local conditions of tectonic stress, which favoured extensive mantle melting and intrusion of high amounts of magmas. In contrast, Sr-Nd isotope ratios are adequate to recognise that rocks with strongest crustal signatures occur in the Western Alps, and are represented by lamproitic dykes and some potassic alkaline intrusive rocks of Valle del Cervo. This may result from different geometry of subducting slab along the Alpine orogeny, which is known to produce different degrees of upper crustal dragging into the mantle, in turn causing variable degrees of both back-arc extension and mantle upraise behind arcs (e.g., Doglioni et al., 1999; Peccerillo and Martinotti, 2006).

## ACKNOWLEDGMENTS

The authors thank Russel Harmon and Riccardo Petrini, University of Trieste, for review of an early version of the manuscript. Research on Alpine magmatism is financially supported by PRIN grant 2008 and by CNR-IGC.

## REFERENCES

Alagna, K. E., Peccerillo, A., Donati, C. (2010). Tertiary to Present evolution of orogenic magmatism in Italy. *Journal of the Virtual Explorer*, 36, paper 18, doi: 10.3809/jvirtex.2009.00233.

Altherr, R., Lugovic, B., Meyer, H. P., and Majer, V., 1995, Early-Miocene postcollisional calc-alkaline magmatism along the easternmost segment of the Periadriatic fault system (Slovenia and Croatia): *Mineralogy and Petrology*, 54, 225–247.

Barth, S., Oberli, F. and Meier, M. (1989). U-Th-Pb systematics of morphologically characterized zircon and allanite: A high resolution isotopic study of the Alpine Rensen pluton (northern Italy). *Earth Planet. Sci. Lett.*, 95, 235-254.

Bellieni, G., Cavazzini, G., Fioretti, A. M., Peccerillo, A. and Poli, G. (1991). Geochemical and isotopic evidence for crystal fractionation, AFC and crustal anatexis in the genesis of the Rensen Plutonic Complex (Eastern Alps, Italy). *Contrib. Mineral. Petrol.*, 92, 21–43.

Bellieni, G., Cavazzini, G., Fioretti, A. M., Peccerillo, A. and Zantedeschi, P. (1996). The Cima di Vila (Zinsnock) Intrusion, Eastern Alps. Evidence for crustal melting, acidmafic magma mingling and wall-rock fluid effects. *Mineral. Petrol.*, 56, 125–146.

Bellieni, G., Fioretti, A. M., Marzoli, A. and Visonà, D. (2010). Permo–Paleogene magmatism in the eastern Alps. Rend. *Fis. Acc. Lincei*, 21, 10.1007/s12210-010-0095-z.

Bellieni, G., Peccerillo, A. and Poli, G. (1981). The Vedrette di Ries (Riesrferner) plutonic complex; petrological and geochemical data bearing on its genesis. *Contrib. Mineral. Petrol.*, 78, 145–156.

Bellieni, G., Peccerillo, A., Poli, G. and Fioretti, A. (1984). The genesis of Late Alpine plutonic bodies of Rensen and Monte Alto (Eastern Alps): Inferences from major and trace element data. *N. Jb. Miner. Abh.*, 149, 209-224.

Berger, A. and Bousquet, R. (2008). Subduction related metamorphism in the Alps: Review of isotopic ages based on petrology and their geodynamic consequences. In: Siegesmund, S., Fügenschuh, B. and Froitzheim, N. (eds), Tectonic Aspects of the Alpine-Dinaride-Carpathian. *Geol. Soc. London Spec. Publ.*, 298, 117–144.

Beltrando, M., Lister, G. S., Rosenbaum, G., Richards, S. and Forster, M. A. (2010). Recognizing episodic lithospheric thinning along a convergent plate margin The example of the Early Oligocene Alps. *Earth Sci. Rev.*, 103, 81-98.

Bigioggero, B., Colombo, A., Del Moro, A., Gregnanin, A., Macera, P. and Tunesi, A. (1994). The Oligocene Valle del Cervo Pluton: an example of shoshonitic magmatism in the Western Italian Alps. *Mem. Soc. Geol. It.*, 46, 409-421.

Borsi, S., Del Moro, A., Sassi, F. P. and Zirpoli, G. (1979). On the age of the Vedrette di Ries (Rieserferner) massif and its geodynamic significance. *Geol. Rundsch.*, 68, 41-60.

Callegari, E. (1983). Geological and petrological aspects of the magmatic activity at Adamello (northern Italy). *Mem. Soc. Geol. It.*, 26, 83–103.

Callegari, E. Cigolini, C. Medeot, O. and D'Antonio, M. (2004). Petrogenesis of calc-alkaline and shoshonitic postcollisional Oligocene volcanics of the Cover Series of the Sesia-Lanzo Zone, Western Italian Alps. *Geodin. Acta*, 17, 1-29.

Carraro, F. and Ferrara, G. (1968). Alpine tonalite at Miagliano, Biella (Zona dioritico-kinzigitica): a preliminary note. *Schweiz. Mineral. Petrogr. Mitt.*, 48, 75-80.

Conticelli S., Guarnieri L., Farinelli A., Mattei M., Avanzinelli R., Bianchini G., Boari E., Tommasini S., Tiepolo M., Prelevi D. and Venturelli G. (2009a) - Trace elements and Sr–Nd–Pb isotopes of K-rich, shoshonitic, and calc-alkaline magmatism of theWestern Mediterranean Region: genesis of ultrapotassic to calc-alkaline magmatic associations in a post-collisional geodynamic setting. *Lithos*, 107, 68–92.

Cortecci, G., Del Moro, A., Leone, G. and Pardini, G. C. (1979). Correlation between strontium and oxygen isotopic compositions of rocks from the Adamello Massif (Northern Italy). *Contrib. Mineral. Petrol.*, 68, 421– 427.

Cumming, G. L., Koeppel, V. and Ferrario, A. (1987). A lead isotope study of the northeastern Ivrea Zone and the adjoining Ceneri zone (N-Italy): evidence for a contaminated subcontinental mantle. *Contrib. Mineral. Petrol.*, 97, 19-30.

Dal Piaz, G. V. (2010). The Italian Alps: a journey across two centuries of Alpine geology. In: Beltrando, M., Peccerillo, A., Mattei, M., Conticelli, S. and Doglioni, C., (dds.), J. *Virtual Expl.*, 36, *paper* 8, *doi:* 10.3809/jvirtex.2010.00234.

Dal Piaz, G.V. and Venturelli, G. (1983). Brevi riflessioni sul magmatismo postofiolitico nel quadro dell'evoluzione spazio-temporale delle Alpi. *Mem. Soc. Geol. It.*, 26, 5-19.

Del Moro, A., Ferrara, G., Tonarini, S. and Callegari, E. (1983a). Rb/Sr systematics on rocks from the Adamello Batholith (Southern Alps). *Mem. Soc. Geol. It.*, 26, 261-284.

Del Moro, A., Pardini, G. C., Quercioli, C., Villa, I. M. and Callegari, E. (1983b). Rb/Sr and K/Ar chronology of Adamello granitoids, *Southern Alps. Mem. Soc. Geol. It.*, 26, 285-299.

Diethelm, K. (1990). Synintrusive basische gänge und endogene xenolithe: magma-minglin in der Bergeller intrusion. *Schweiz. Mineral. Petrogr. Mitt.*, 70, 247-264.

Doglioni, C., Harabaglia, P., Merlini, S., Mongelli, F., Peccerillo, A. and Piromallo, C. (1999). Orogens and slabs vs. their direction of subduction. *Earth Sci. Review*, 45, 167-208.

Dupuy, C., Dostal, J. and Fratta, M. (1982). Geochemistry of the Adamello Massif (Northern Italy). *Contrib. Mineral. Petrol.*, 80, 41-48.

Handy, M. R., Schmid, S., Bousquet, R., Kissling, E. and Bernoulli, D. (2010). Reconciling plate-tectonic reconstructions of Alpine Tethys with the geological-geophysical record of spreading and subduction in the Alps. *Earth Sci. Rev.*, 10.1016/j.earscirev.2010.06.002.

Harmon, R. S., Halliday, A. N., Clayburn, J. A. P., and Stephens, W. E. (1984). Chemical and isotopic systematics of the Caledonian intrusions of Scotland and Northern England: *A guide to magma source region and magma-crust interaction*. Phil. Trans. Royal Society London, A310:709-742.

Harmon, R. S. and Hoefs, J. (1995). Oxygen-isotope heterogeneity of the mantle deduced from global O-18 systematics of basalts from different geotectonic settings. *Contrib. Mineral. Petrol.*, 120, 75-114.

Kagami, H., Ulmer, P., Hansmann, W., Dietrich, V. and Steiger, R. H. (1991). Nd–Sr isotopic and geochemical characteristics of the southern Adamello (Northern Italy) intrusives: implications for crustal versus mantle origin. *J. Geophys. Res.*, 96 (B9), 14331–14346.

Kessel, I. R., Ulmer, P. and Pettke, T. (2005) Trace Element Signatures Of Subduction-Zone Fluids, Melts And Supercritical Liquids At 120-180 Km Depth. *Nature,* 437, 724-727.

Kovács, I., Csontos, L., Szabó, Cs., Bali, E., Falus, G., Benedek, K. and Zajacz, Z. ( 2007). Paleogene–early Miocene igneous rocks and geodynamics of the Alpine-Carpathian-Pannonian-Dinaric region: An integrated approach, in Beccaluva, L., Bianchini, G., and Wilson, M., eds., Cenozoic Volcanism in the Mediterranean Area. *Geol. Soc. Am. Special Paper*, 418, 93–112.

Liati, A., Gebauer, D. And Fanning, C. M. (2000). U–Pb SHRIMP dating of zircon from the Novate Granite (Bergell, Central Alps): evidence for Oligocene–Miocene magmatism, Jurassic/Cretaceous continental rifting and opening of the Valais Trough. *Schweiz. Mineral. Petrogr. Mitt.*, 80, 305–316.

Lustrino, M. and Wilson, M. (2007). The circum-Mediterranean anorogenic Cenozoic igneous province. *Earth Sci. Rev.*, 81, 1-65.

Macera, P., Gasperini, D., Ranalli, G. and Mahatsente, R. (2008). Slab detachment and mantle plume upwelling in subduction zones: An example from the Italian southeastern Alps volcanism. *J. Geodynam.*, 45, 32-48.

Pamic, J. and Palinkas, L. (2000). Petrology and geochemistry of Paleogene tonalites from the easternmost parts of the Periadriatic Zone. *Mineral. Petrol.*, 70, 121-141.

Peccerillo, A. And Martinotti, G. (2006). The Western Mediterranean lamproitic magmatism: origin and geodynamic significance. *Terra Nova* 18, 109–117.

Pinarelli, L., Bergomi, M.A., Boriani, L. And Giobbi E. (2008). Pre-metamorphic melt infiltration in metasediments: geochemical, isotopic (Sr, Nd, and Pb), and field evidence from Serie dei Laghi (Southern Alps, Italy). *Mineral. Petrol.*, 93, 213-242.

Pinarelli, L., Boriani, A. and Del Moro, A. (1993). The Pb isotopic systematics during crustal contamination of subcrustal magmas: the Hercynian magmatism in the Serie dei Laghi (Southern Alps, Italy). *Lithos,* 31, 51-61.

Rosenbaum, G. and Lister, G. S. (2005). The Western Alps from the Jurassic to Oligocene: spatio-temporal constraints and evolutionary reconstructions. *Earth Sci. Rev.*, 69, 281-306.

Rosenberg, C. L. (2004). Shear zones and magma ascent: A model based on a review of the Tertiary magmatism in the Alps. *Tectonics,* 23, TC3002, 10.1029/2003TC001526.

Ruffini, R., Polino, R., Callegari, E., Hunziker, J.C. and Pfeifer, H. R. (1997). Volcanic clast rich turbidites of the Taveyanne Sandstones from the Thônes syncline (Savoie, France): records for a Tertiary postcollisional volcanism. *Schweiz. Mineral. Petrogr. Mitt.*, 77 (2), 161–174.

Sassi, F. P., Cavazzini, G.C. and Visonà, D. (1985). Radiometric geochronology in the eastern Alps: results and problems. Rend. Soc. *It. Min. Petrol.*, 40, 187-224.

Schmid, S. M., Fügenschuh, B., Kissling, E. and Schuster, R. (2004). Tectonic map and overall architecture of the Alpine orogen. *Eclogae Geol. Helv.,* 97, 93-117.

Stampfli, G. M. and Hochard, C. (2009). Plate tectonics of the Alpine realm. In: Murphy, J. B., Keppie, J. D. and Hynes, A. J. (eds.) Ancient orogens and modern analogues. *Geol. Soc. London Spec. Publ.*, 327, 89-111.

Stracke, A, Hofmann, A. W. and Hart, S. R. (2005). *FOZO, HIMU, and the rest of the mantle zoo.* G3, 6 (5), 1-20.

Sun, S. and McDonough, W. F. (1989). Chemical and isotopic systematics of oceanic basalts: implications for mantle composition and processes. In: Saunders A. D., Norry M. G. (eds) Magmatism in the Ocean Basins. *Geol. Soc. Amer. Spec. Publ.*, 42, 313-345.

Tiepolo, M. and Tribuzio, R. (2005). Slab-melting during Alpine orogeny: evidence from mafic cumulates of the Adamello batholith (Central Alps, Italy). *Chem. Geol.*, 216, 271-288.

Tiepolo, M., Tribuzio, R. and Vannucci, R. (2002). The composition of mantle-derived melts developed during the Alpine continental collision. *Contrib. Mineral. Petrol.*, 144, 1-15.

Ulmer, P., Callegari, E. and Sonderegger, U. C. (1983). Genesis of the mafic and ultramafic rocks and their genetical relations to the tonalitic–trondhjemitic granitoids of the southern part of the Adamello Batholith (Northern Italy). *Mem. Soc. Geol. It.,* 26, 171–222.

van Marcke de Lummen, G. and Vander Auwera, J. (1990). Petrogenesis of the Traversella diorite (Piemonte, Italy): a major and trace elements and isotopic (O, Sr) modelling. *Lithos,* 24, 121-136.

Venturelli, G., Thorpe, R. S., Dal Piaz, G. V., Del Moro, A. and Potts, P. J. (1984). Petrogenesis of calc-alkaline, shoshonitic and associated ultrapotassic Oligocene volcanic rocks from the Northwestern Alps, Italy. *Contrib. Mineral. Petrol.*, 86, 209–220.

von Blanckenburg, F., Kagami, H., Deutsch, A., Oberli, F., Meier, M., Wiedenbeck, M., Barth, S. & Fischer, H. (1998). The origin of Alpine plutons along the Periadriatic Lineament. *Schweiz. Mineral. Petrogr. Mitt.*, 78, 55-66.

von Blanckenburg, F., Früh-Green, G., Diethelm, K. and Stille, P. (1992). Nd-, Sr-, O-isotopic and chemical evidence for a two-stage contamination history of mantle magma in the Central-Alpine Bergell intrusion. *Contrib. Mineral.Petrol.,* 110, 33–45.

Voshage, H., Hofmann, A. W., Mazzucchelli, M., Rivalenti, G., Sinigoi, S., Raczek, I. and Demarchi, G. (1990). Isotopic evidence from the Ivrea Zone for a hybrid lower crust formed by magmatic underplating. *Nature*, 347, 731-736.

Voshage, H., Hunziker, J. C., Hofmann, A. W. and Zingg, A. (1987). A Nd and Sr isotopic study of the Ivrea zone, Southern Alps, N-Italy. *Contrib. Mineral. Petrol.*, 97, 31-42.

Zartman, R. E. and Doe, B. R. (1981). Plumbotectonics - *The model. Tectonophysics*, 75, 135-162.


*Reviewed by* Riccardo Petrini, University of Trieste, Italy, and Russel Harmon, International Research Office of the US Army, USA.

In: Granite  
Editors: Miroslava Blasik and Bogdashka Hanika

ISBN: 978-1-62081-566-3  
© 2012 Nova Science Publishers, Inc.

*Chapter 4*

# CONTRASTING PETROLOGICAL ATTRIBUTES OF GRANITES IN NIGERIA

### *Ademuyiwa Adetunji**

Department of Applied Geology, School of Earth and Mineral Sciences,  
Federal University of Technology, Akure, Ondo State, Nigeria

## ABSTRACT

In Nigeria, granites were emplaced at two different geological ages in contrasting tectonic regimes. One was emplaced during the Pan-African orogeny in the Neoproterozoic (ca. 600 Ma) while the other was associated with the Mesozoic anorogenic tectonic event which was localized to the northcentral Nigeria. The older of this is referred to as the Pan-African Granites or the Older Granites while the younger is often referred to as the Younger Granites.

The Pan-African Granites represent the most prominent features of the Pan-African orogeny and are essentially pre-, syn- and post-tectonic. They occur as stocks, bosses and batholiths, and are widely distributed in the Basement Complex of Nigeria. The Pan-African Granites with other plutonic igneous rocks of the same age range are collectively called Pan-African Granitoids. The granitoids range in composition from diorites and tonalites through granodiorites to true granites and syenite. Charnockites, pegmatites and aplites are also associated with them. The Pan-African Granites constitute a major petrological unit of the Basement Complex of Nigeria. They are notable for lack of any mineralization. However, certain categories of the associated pegmatites contain economic quantity and grade of many rare metals: Sn, Nb, Ta, Li etc. and host of gemstones: tourmaline, beryl, garnet and industrial minerals like feldspars, muscovite and quartz.

The Younger Granites occur as groups of ring dykes called ring complexes. They constitute parts of a larger province of mid-plate anorogenic alkaline magmatism of the Mesozoic period. They occupy a zone of about 200 km wide and extends from the Benue Valley in the south to about 320 km northwards within the country. Outside the country, the zone extends to a distance of about 1,600 km into the Republic of Niger. The ages of the granites running from the most northerly Niger range from about 470 ± 5 Ma in Air

---

* E-mail: aamutiwa@yahoo.co.uk, aadetunji@futa.edu.ng. Tel: +234(0)8025915452.

through 310 Ma in Zinda to about 141 Ma in Afu at the southernmost in Nigeria. The Younger Granites are essentially high-level intrusives and mostly peralkaline with subordinate peraluminous and metaluminous. More than 50 ring complexes have been identified. The diameters of the complexes vary from < 2 to > 25 km and individual massifs varying from about 1,000 km$^2$ to < 1 km$^2$. Four major petrographic groups of the Younger Granites are distinguished: hornblende-pyroxene-fayalite granites, hornblende-biotite-granites, biotite granites and riebeckite granites. Other associated igneous rocks are syenite, gabbros, rhyolites, dolerites, basalts, agglomerate and tuffs.

The albitized riebeckite granites and albitized biotite granites are known to contain Nb, Sn, U, Li and PO$_4$ mineralization. So also, the associated minor pegmatites, pegmatitic granites, greisens and quartz veins contain Sn, Zn, Be, Pb, Cu, W and Th mineralization. The alluvial and eluvial deposits of Sn and Nb derived from these mineralized Younger Granites largely supported the economy of Nigeria up to the late 1950's when crude oil was discovered in commercial quantity.

**Keywords**: Nigeria, Granites, Pan-African, Mesozoic, Ring complexes

# INTRODUCTION

Situated between the West African and Congo Cratons, Nigeria occupies part of the areas that were reactivated during the Pan-African orogeny (600±150 Ma). The total land mass of Nigeria is about 923,768 Km$^2$ and is underlain in nearly equal proportions by both sedimentary and crystalline rocks (Figure 1). The evolution of the sedimentary basins postdated the Pan-African orogeny. There are seven major basins: Sokoto, Bornu or Chad, Bida, Benue trough, Dahomey or Benin, Anambra and highly petroliferous Niger Delta Basins. The sediments in these basins aged from the Cretaceous to Recent.

The crystalline rocks are divided into three (3) groups:

- The Basement Complex
- The Younger Granites
- The Tertiary to Recent Volcanics

The basement rocks are all Precambrian in age. The Basement Complex has been grouped into six petrologic units (Rahaman, 1988):

1. Migmatite-gneiss-quartzite complex;
2. Metasedimentary and metaigneous rocks (schist belts);
3. Charnockitic-dioritic-gabbroic rocks;
4. The Older Granite (Pan-African Granite) suite;
5. Metamorphosed to unmetamorphosed calc-alkaline volcanic and hypabyssal rocks; and
6. Unmetamorphosed dolerite dykes, basic dykes and syenite dykes (acid and basic dykes).

The above petrologic units are based on detailed geological observations. However, in recent literatures only four petrologic units are distinguished (e.g. Obaje, 2009). These are units 1, 2, 4 and 6 of Rahaman (1988). These literatures considered unit 3 as part of unit 4 and unit 6 as part of unit 5. The six-fold classification has more merits than the four-fold classification as it takes into cognizance detailed petrographic observations. For detailed review of the Basement Complex of Nigeria, interested reader can consult Jones and Hockey (1964), Rahaman (1976, 1988); Rahaman and Ocan (1978); Dada (2006) and Obaje (2009). The basement rocks are believed to be the results of at least four cycles of deformation, metamorphism and remobilization corresponding to four different orogenic events: Liberian (2,700 Ma), Eburnean (2,000 Ma), Kibaran (1,100 Ma) and Pan-African (600 ± 150 Ma). The last has largely overprinted the earlier ones.

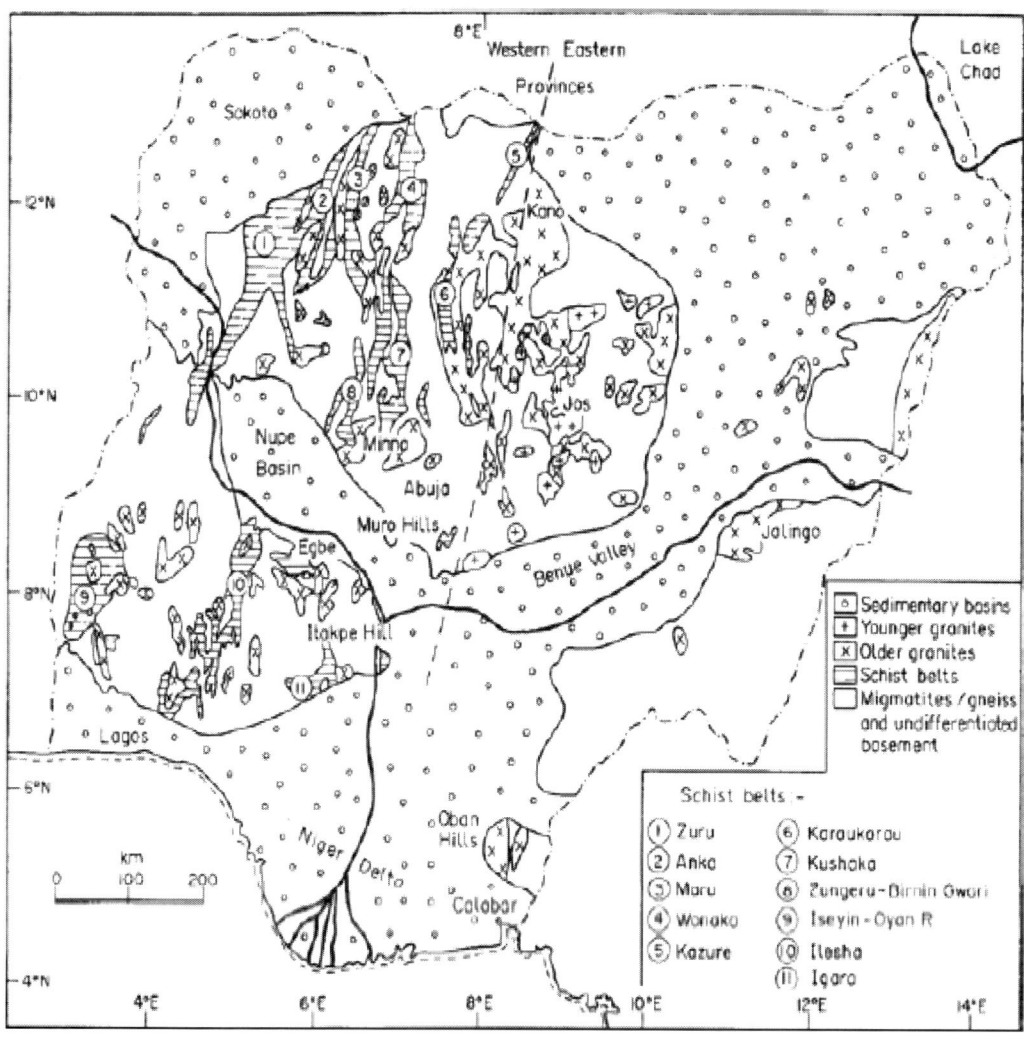

(After Woakes et al., 1987).

Figure 1. Generalized geological map of Nigeria.

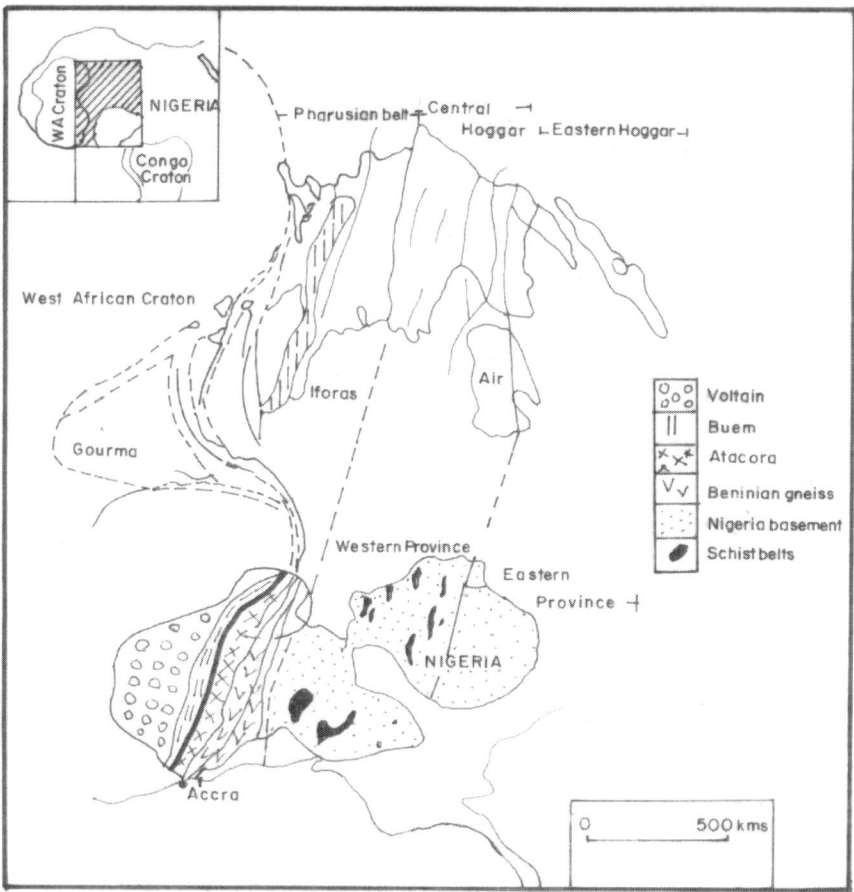

(After Caby et al., 1981).

Figure 2. Regional geological setting of Trans Saharan fold belt.

## Regional Geology and Tectonic Evolution

The entire Pan-African mobile belt east of the West African and northwest of the Congo cratons is referred to as the Trans Saharan fold belt (Caby, 1989). The fold belt extends from the Hoggar (Algeria) in the north through Air, Adrar des Iforas and Gourma to Ghana, Togo, Benin, Nigeria and Cameroon in the south (Figure 2). The belt is made up of several continental blocks probably amalgamated during oblique collision (Ajibade and Fitches, 1988). The southern portion of the Trans Saharan fold belt is referred to as the Dahomeyide (Affaton et al., 1991). From the southeastern margin of the West African craton eastwards, Petters (1991) and Affaton et al. (1991) recognized three major tectonic domains. These are the Voltain foreland basin with sedimentary sequences, the Beninian thrust and fold belt, and the Nigerian high-grade migmatite-gneiss terrane. The Voltain Basin has sedimentary sequences that have progressively been folded and metamorphosed eastwards. The rocks in the basin rest unconformably on the older rocks. The Beninian fold belt has two main structural units called Buem and Atacora (Petters, 1991). The Buem is southwestward dipping and largely consists of unmetamorphosed sandstones, shales and mudstones. Slices of dismembered ophiolites have been mapped here. The Buem unit is well developed in Ghana. The Atacora unit consists of quartzite, mica schist, phyllite, basaltic volcanics and marble. The internal unit attained a high-pressure metamorphic grade in southern part of Togo where kyanite-bearing eclogite has been mapped (Petters, 1991).

The Nigerian province consists of two main areas: the Beninian gneisses and migmatite-gneiss terrane of Nigeria (Caby, 1989). The Beninian gneisses are high-grade amphibolite facies predominantly quartzo-feldspathic and anatectic migmatitic granitoids. The migmatite-gneiss complex of Nigeria consists of Archean polycyclic grey gneisses of granodioritic to tonalitic composition. There is a wide spread presence of Proterozoic cover now represented by metasediments. These have been interpreted as the Schist belts (Ajibade, 1976).

Most workers in this area appear to agree that plate tectonics models could be used to explain the evolution of this Pan-African mobile belt (Burke and Dewey, 1972; and Affaton et al., 1991). Burke and Dewey (1972, 1973) suggested that the Pan-African mobile belt evolved from the closure of an Atlantic-Type Ocean by continent-continent collision between passive western continent (Birrimia) and an active eastern continent (Dahomea). Continental fragmentation commenced ca. 1100 Ma and by ca. 900 ± 100 Ma a paleo-ocean had developed east of the West African craton. This led to the development of a passive western continental margin (Voltain Basin) (Affaton et al., 1991). The closure of this ocean began with subduction processes operating along an eastward dipping Benioff Zone. This was marked by widespread calc-alkaline volcanisms that are similar to those of the modern day island arcs and active continental margins (Affaton et al., 1991). Behind the possible Pan-African arc were several often fault bounded Pan-African volcano-sedimentary sequences which latter evolved to schist belts (Affaton et al., 1991). Closure occurred along the eastern margin of the West African craton.

## The Pan-African Granites

The Pan-African Granites occur in all the three main areas of Nigeria where Basement Complex rocks outcrop, namely:

- Southwestern Nigeria,
- Northcentral Nigeria, and
- Southeastern border of Nigeria with Cameroon.

In Nigeria, the Pan-African Granites are regarded as Older Granites. Older Granite was coined by Falconer (1911), King and de Swardt (1949) and de Swardt (1953). Older Granites was initially used by these authors to distinguish these orogenic related granites from the Mesozoic anorogenic, high level granites of the northcentral Nigeria. Neoproterozoic (ca. 600 Ma) U-Pb, Rb-Sr and K-Ar ages have been reported from the Older Granites and hence the term Pan-African Granites is preferred to Older Granites (Van Breemen et al., 1977; Dada et al., 1989 and Rahaman et al., 1991). The granites are associated with other intrusive rocks of similar ages, hence they are collectively called Pan-African Granitoids. The granitoids consist of granites, quartz monzonite, granodiorites, syenites, charnockites and pegmatite. Both Pan-African and Older Granites are still interchangeably used in literatures of geology of Nigeria.

## Field Occurrence and Geological Features of the Pan-African Granites

The Pan-African Granites are the most prominent features of the Pan-African orogeny in Nigeria. They account for about 70% of the total basement rocks in the northcentral Nigeria and about 20 % of the surface area in the southwestern Nigeria. The granites outcrop as distinct plutons usually of stock, boss and batholithic dimensions. In many places where the granites outcrop, they form prominent landforms like inselbergs with smooth exfoliation surfaces. In other areas of the country, the rocks outcrop as hummocky hills, whale and turtle backs (Rahaman et al., 2005). Discrete bodies, which are several kilometers across are also present within the basement.

The Pan-African Granites intruded migmatite-gneiss-quartzite complex and schists. They are generally elongated parallel to the regional structure i.e. N-S trend (e.g. Ferre et al., 1995; and Beka and Ukaegbu, 2008). Field relationships with respect to the actual contacts between the individual Pan-African Granites and between the granites and other rock types are difficult to determine owing to the poor exposures in most critical areas (Rahaman et al., 2005). However, sharp and intrusive gradational and replacive contacts exist between the granites themselves on one hand and between the granites and country rocks on the other. There are textural and sometimes lithological variations in the granites from the margins with country rocks inwards. Xenoliths of the country rocks are common in the granites.

Rahaman (1976; 1988) used textures and fabrics to classify Pan-African Granites into seven distinct types:

1. migmatitic granite;
2. granite gneiss;
3. early pegmatites and fine-grained granites;
4. homogeneous medium to coarse-grained granites;
5. medium to coarse grained porphyritic granite;
6. slightly deformed pegmatites, and aplites dykes and veins; and
7. undeformed pegmatites, two mica granites and quartz veins.

The most abundant of these types are the coarse-grained porphyritic variety. There have been attempts to classify the Pan-African Granites on the basis of tectonic features and timing of emplacement during the Pan-African orogeny. Jones and Hockey (1964) recognized three main groups, namely the early phase comprising granodiorites, main phase porphyritic granites and late phase even-grained granites, and pegmatite and aplite dykes. These phases are respectively described as pre-tectonic, syn-tectonic, post-tectonic (or late tectonic). Both pre-tectonic and syn-tectonic granites were affected by crystal plastic deformation with penetrative fabrics. McCurry and Wright (1977) considered plutons emplaced into the gneisses as syn-kinematic while those emplaced into the metasediments as post-kinematic. Rahaman (1973) equated late-kinematic and post-kinematic granites to the main and late phases of Jones and Hockey (1964).

The classification on the basis of tectonic setting using geochemistry based on Pearce et al. (1984) has been attempted for some individual plutons. For example, Okonkwo and Winchester (2004) inferred that undeformed granites around Jebba, southwestern Nigeria have attributes similar to within plate granite while the deformed variety is similar to rocks of volcanic arc environment. In the same vein, Ukwang and Ekwueme (2009) classified granites in southeastern Nigeria as arc and collisional types. The different tectonic settings of emplacement may be largely due to long span of Pan-African orogeny (600±150 Ma).

## Structural Features and Petrography of the Pan-African Granites

The Pan-African Granites are characterized by pronounced foliation. This foliation has long been recognized by Trunswell and Cope (1963). The foliation ranges from crude to strong and is defined by parallel alignment of longer axes of feldspar phenocrysts. Ferre et al. (1995) described this type of foliation in Solli pluton (northcentral Nigeria) as magmatic. Equally, biotite and hornblende show this alignment in many cases. Flow banding and structures have been recognized in many plutons. The orthogneissic banding in the granites is interpreted as solid state deformation (Ferre et al., 1995). Generally, a north-south alignment has been noticed all over the country (e.g. Rahaman, 1973, Ferre et al., 1995). Despite this general trend, there are phenocrysts of feldspars that are haphazardly arranged in some of the plutons (Rahaman, 1976). The flattening of some of the mafic enclaves in the granites point to a limited strain of the host granites (Ferre et al., 1995).

The Pan-African Granites (*sensu stricto*) have simple mineralogy as they all contain essentially quartz, potash feldspar which is mainly microcline/microcline-perthite and plagioclase (albite to oligoclase). The other minerals are biotite and /or hornblende. The accessories include apatite, zircon, allanite, sphene and opaque ores (Olarewaju, 1981). However, the two mica granites contain both muscovite and biotite. In the homogeneous highly peraluminous rocks, blebs of garnet can be present. Okonkwo and Winchester (2004) reported the presence of secondary epidote and chlorite in coarse porphyritic rocks in Jebba area, southwestern Nigeria.

According to Rahaman et al. (2005), the features of petrographic interest include widespread myrmekitic intergrowth, hornblende mantling biotite, zoned plagioclase, undulose extinction in quartz and saussuritisation of plagioclase and seriticization of both feldspars. Phenocrysts of microcline-perthite are ubiquitous while in some cases orthoclase is present.

## Geochemistry

Geochemical studies show the granites to contain high amount of alkalis and are slightly peraluminous (e.g.Odeyemi, 1977; Ajibade, 1980; Olarewaju, 1981; Okonkwo and Winchester, 2004). According to Rahaman (1988), the rocks show two trends in the AFM diagram. Most of the rocks plot in the field that corresponds to plutonic calc-alkaline rocks and second trend shows Fe-enriched fractionation. Using the plot of $SiO_2$ vs. $K_2O$ of Rickwood (1989) and FeO/FeO+MgO vs $SiO_2$ (Frost etal. 2001), Okonkwo and Winchester (2004) show that the granites of Jebba area are essentially high-K, calc-alkaline, Fe-enriched type. However, Beka and Ukaegbu (2008) recognized iron depleted calc-alkaline, high potassic granites in the southeastern Nigeria. The Rb/Sr isotopic studies (Van Breemen et al., 1977; Ajibade, 1980; Matheis and Cean-Vachette, 1983) show that the initial Sr ratio to be about 0.708. This figure is higher than the mantle average (0.702) and lower than the crustal average (0.75). Based on this observation, the Pan-African Granites are interpreted to be derived from the anatexis of low Rb/Sr source rocks or crustal contamination of mantle-derived magma. Olarewaju (1981) and Beka and Ukaegbu (2008) reached similar conclusion based on trace element and REE data respectively.

## Emplacement Mechanism

Owing to paucity of field and structural data, there is no universal emplacement mechanism for the Pan-African Granites in Nigeria. This is complicated by rare exposures of contacts with the country rocks. However, geochemical and isotopic studies have shown that these granites have either I-type, S-type or I- and S-type affinity (Ajibade et al., 1987). The most detailed structural study of the granite is by Ferre et al. (1995). The authors used some physical measurements, internal fabrics and microstructural elements to infer the emplacement mechanism of Solli pluton (northcentral Nigeria) as through dextral tear fault system. The study provides petrostuctural data that illustrate the case of strike-slip pluton emplacement with passive incremental growth through feeder plutons. Rahaman et al. (2005) argued that both passive and active emplacement have been proposed for the Pan-African Granites. For the plutons emplaced through forceful mechanisms, evidence is provided by folding and fracturing of rocks around the plutons. This appears to be the mechanism of emplacement of some plutons in the southwestern Nigeria. Mullan (1979) observed that the final emplacement of granites in central Nigeria was through the incremental growth.

Clemens (1998) observed that evidences generally favour dyke ascent of granitic magma. Modeling has shown that the ascent of granitic magma is rapid and thermally efficient. Granitic dykes are common, this indicates that it is reasonable to consider ascent of granitic magmas through fracture systems (Rahaman et al., 2005). This concept of feeder plutons is consistent with Ferre et al. (1995).

It is thus plausible that the emplacement of some of the Pan-African Granites have been by dyking mechanism (Rahaman et al., 2005). However, it is essential to look for the field evidence that support any the mechanism of emplacement of a particular body.

## Mineralization

The Pan-African Granites are generally notable for lack of mineralization. However, Odigi (1986) in his study of the Pan-African Granites of Oban Massif in the eastern Nigeria noticed that some of the granites are enriched in tin, topaz, tourmaline, fluorite and apatite. This was attributed to influx of some hydrothermal fluids, which eventually caused wall rock alteration and introduction of rare elements. This appears to be the only reported case of evidence of mineralization in the Pan-African Granites.

However, certain categories of associated pegmatites have been known to contain economic grade and quantity of rare elements bearing minerals. Tin, Nb, Ta, Li, W etc. and host of gemstones: tourmaline, beryl, garnet and industrial minerals such as feldspars, muscovite and quartz have been recovered from these pegmatites. These mineralized pegmatites are widespread in the entire Basement Complex of Nigeria.

## The Younger Granites

The term Younger Granites were first used by Falconer (1911) to distinguish them from the Older Pan-African Granites. The Younger Granites of Nigeria are restricted to the Jos plateau and surrounding areas. Their emplacement is not related to any orogenic activity, hence they are regarded as anorogenic. The granites occur in a zone of about 200 km wide and extends from the Benue Valley in the south to about 320 km northwards within the country (Ekwueme, 1993). Regionally, it extends to a distance of about 1,600 km into the Republic of Niger in the north (Obaje, 2009) (Figures 3 and 4). The chain according to Turner (1976) corresponds to the N-S belt of the Pan-African orogeny and forms the northerly continuation of the continental margin of South Africa. The Younger Granites constitute one of the world's best mid plate magmatism (Kinnnaird et al., 1985). Because of this and its structural architecture and mineralization, it has attracted much studies than the Pan-African Granites.

## Field Occurrence of the Younger Granites in Nigeria

The Younger Granites occur as groups of ring dykes called ring complexes (Turner, 1976; Bowden and Turner, 1974; Obaje, 2009; Ekwueme, 1993). The ring complexes cover a total area of about 7,500 $Km^2$ with individual massifs varying from about 1,000 $Km^2$ to less than 1 $Km^2$. More than fifty (50) of such ring complexes have been mapped. Some complexes have concentric patterns, indicating that the activity was confined to one area, but others have overlapping rings. This shows that the centre of magmatic activity migrated with time.

The majority of the complexes are between 100 and 250 $Km^2$ with circular or elliptical outlines. Each of the complexes began as chains of volcanoes (Bowden, 1985). The complexes are made up of mostly granites (more than 95%) ring dykes. Each dyke varies in plan from polygonal to circular or crescent. The intermediate and mafic components of the ring complexes are less 5% (Obaje, 2009).

## Petrography of the Younger Granites

Ekwueme (1993) gave the summary of the four major petrographic groups of the Younger Granites as follows:

1. Hornblende-pyroxene-fayalite granites: This is essentially greenish with porphyritic texture. The phenocrysts are quartz and feldspar. These ring dykes have chilled zones near their margins and resemble quartz porphyries.
2. Hornblende-biotite-granites: These are granular rocks and outcrop in all the complexes. They are characterized by mafic clots and needles of apatite.
3. Biotite-granites: This petrographic unit is characterized by granular texture. Quartz and biotite often cluster into rounded aggregate. Accessory fluorite, cassiterite and columbite are present in the fine-grained albite bearing variety.

4. Riebeckite granites: This is less well represented but form significant body in the Sara-Feir complex (Figure 4).

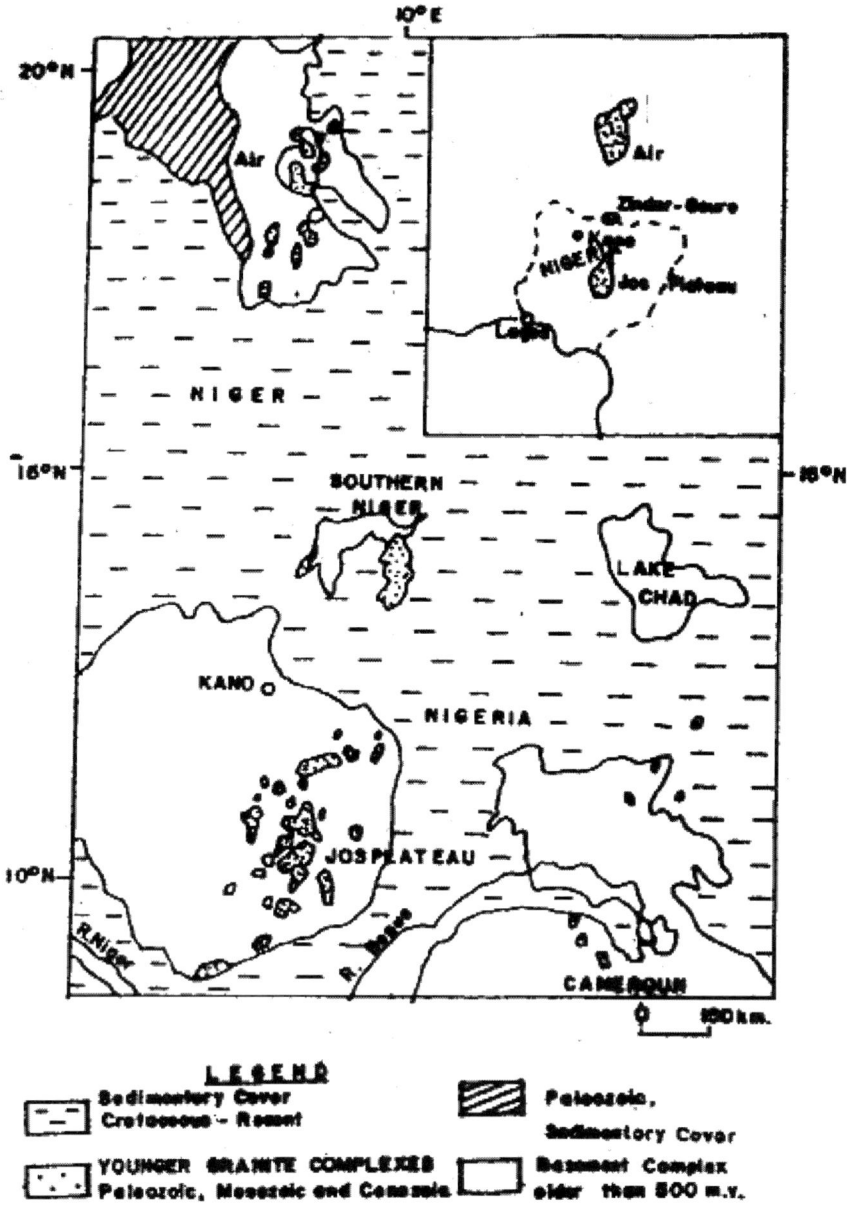

(Modified from Rahaman et al., 1984).

Figure 3. Location map showing the distribution of the Younger Granites ring complexes.

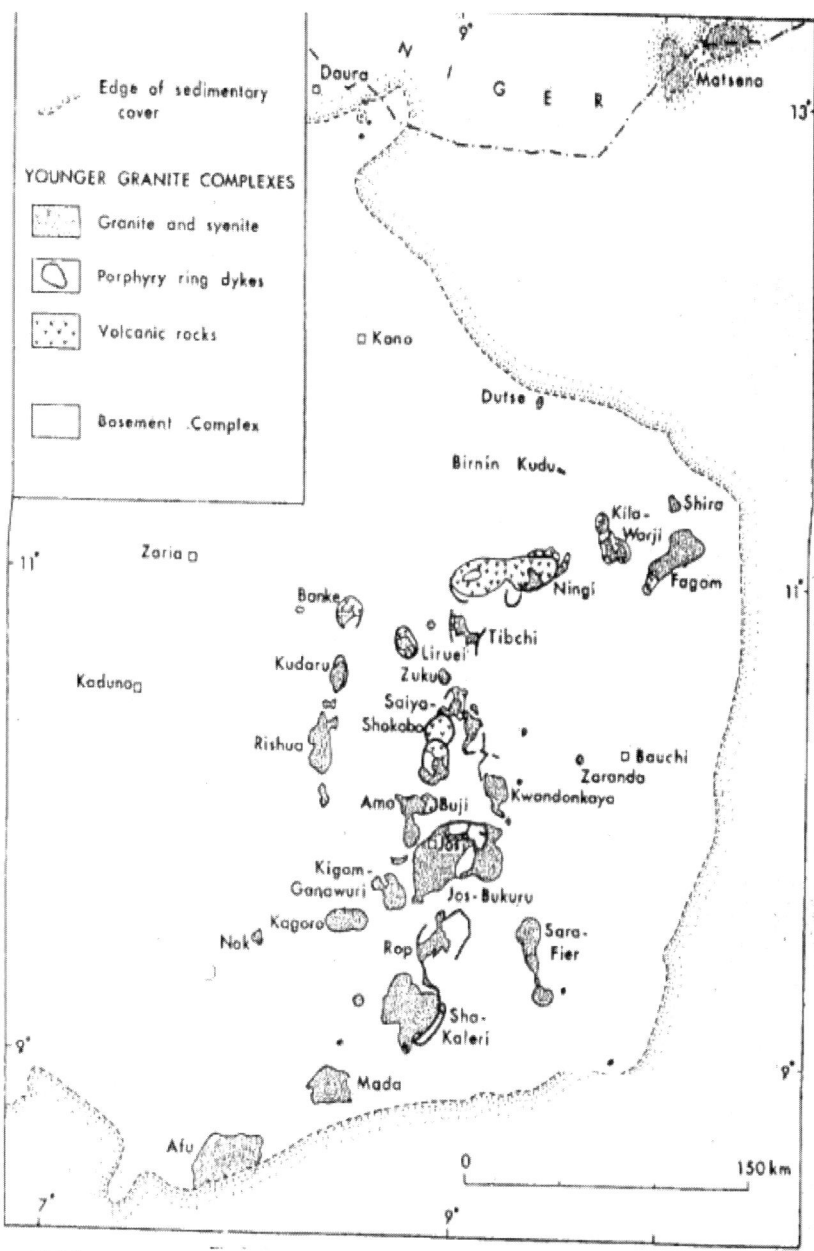

(After Turner, 1976)

Figure 4. Distribution of the Younger Granites ring complexes within Nigeria.

There are other minor varieties that are named after the characteristic amphiboles and pyroxenes they contain. These minerals include arfvedsonite, aegerine, hastingsite and hedenbergite. In addition, there are albitized varieties: albite-biotite granite and albite-riebeckite-afvedsonite granites. The granites are described as peralkaline with subordinate peraluminous and metaluminous suites. Some of the peraluminous varieties contain garnet as exemplified by part of Mada complex. Other rocks that are associated with the granites are rhyolites (early and late), gabbros, syenite, dolerites, basalts, ignimbrites, tuffs and agglomerates.

## Geochemistry of the Younger Granites

Chemically, the biotite granites have been shown to be peraluminous with $Fe_2O_3$ lower than that of peralkaline granite, but slightly higher $Al_2O_3$ in the peralkaline granite (Bowden et al., 1976). The Sn-rich granites are low in MgO, CaO and poor in FeO. The ratio of Na/K varies greatly. The biotite granites with slight excess Na contain assemblage of rare elements like Nb and Sn.

Cesium is geochemically close to Rb, Tl and Rb and generally concentrates in the K minerals. The Younger Granites generally have low Cs but slight enrichment in tourmaline bearing variety. Strontium, Nd and Pb isotopic studies on the Younger Granite samples from Ririwai suggest that the sources of the magma and ore mineralization lies partly in the Precambrian continental crust and partly in the mantle (Kinnaird et al., 1985). The REE chondrite normalized curve for some riebeckite and biotite granites show horizontal or slightly decline line with a negative slope (Bowden et al., 1976). This suggests a marked enrichment in the heavier REE over the lighter ones. Equally, the rocks show marked Eu depletion. Due to the REE distribution pattern, the authors suggested that REE might be used as exploratory tool in the Younger Granites province. The Sr initial ratio for the granites was estimated to be $0.721 \pm 0.004$ (Bowden et al., 1976). From this isotopic study, these authors concluded that the granites might have a common origin.

## Geochronology of the Younger Granites

Fifteen of the over fifty complexes have been isotopically dated and they show a decreasing age southwards. In the Republic of Niger, the Air complex yielded $470\pm5$ Ma, while Zinda gave 310 Ma (cf. Ekwueme, 1993). Rahaman et al. (1984) obtained the following ages starting from southern Niger to southernmost part of the ring complexes in Nigeria: Goure (264 Ma) and Matsena (258 Ma) in Niger, Dutse (213 Ma), Fagam (191 Ma), Ningi (183 Ma), Banke (173 Ma), Zaranda (186 Ma), Jos (161 Ma), Pankshin (151 Ma), Mada (147 Ma) and Afu (141 Ma) all in Nigeria. This chain appears to be N-S trend in the Niger (Turner, 1976) but with a clear NE alignment in Nigeria (Obaje, 2009). Aeromagnetic anomalies suggest a series of buried NE-SW lineaments of incipient rifts controlled disposition of individual complexes (Ajakaiye, 1983). Rahaman et al. (1984) substantiated that major local migrations of magmatic activity were concentrated along at least two ENE-WSW zones (Figures 3 and 4). They noted that the age pattern suggest that contemporaneous anorogenic magmatic activity may have been locally derived from several simultaneous high level magma chambers connected to a common deeper source.

## Evolution and Origin of the Younger Granites

Three stage evolutionary processes have been proposed for the development of the Younger Granites (ring complexes) of Nigeria (Turner, 1976):

1. Early volcanic stage: This is marked by building up of large rhyolite volcano in two phases (early and late rhyolite phases). Before the end of this stage, large amount of

magma has accumulated in subvolcanic reservoir about 5 km beneath the surface. The early rhyolite is banded and the succession includes tuffs, agglomerates, and occasional basalts and trachytes. The early rhyolite is overlain by late massive porphyritic rhyolite.

2. Caldera and ring dyke stage: The early volcanic stage probably ended with rapid eruption of ash flow tuffs, this partly emptied the magma reservoir and led to the collapse of the centre of the volcanic structure within a ring fault. The magma rose along this fault and crystallized to form granite porphyry or extruded into the caldera to form late rhyolite. The rocks of this stage are all porphyritic.

3. Intrusive stage:. This is the waning phase of igneous activity and it was marked by crystallization of smaller granite bodies at increasingly deeper levels. The magma in this stage evolved and crystallized at lower temperatures with lower assemblages than zones of the outer ring dykes. Aegerine-arfvedsonite and hastingsite-biotite granites were emplaced during this stage.

Based on the general southwards decrease in age of the ring complexes, Bowden and Bennett (1977) noted that the main phase of acid magmatism commenced in the Triassic and migrated in southerly direction until at least to the close of the Jurassic. The cessation of magmatism was synchronous with the initiation of the opening of the South Atlantic. The rifting and separation of Africa from South America was guided by the structural trends in the basement (Turner, 1976). He postulated that the Younger Granites lie on the extension of the ancient ridge structure that did not develop a central rift valley but developed into a zone of migratory magmatic activity. Aeromagnetic studies confirmed the presence of this buried fracture zone (Ajakaiye, 1983).

Some workers have advocated upper mantle origin of the granites (Ekwueme, 1993). This is based on limited occurrence of xenoliths, gabbros, basalts and hybrid rocks. However, Bowden and Turner (1974) suggested that the non peralkaline rocks evolved from the lower crust. Hence, lower crust may have been the source of the Younger Granites. Geochemical evidence indicates that sialic crust, lower crust and the upper mantle may all have contributed to the evolution of the Younger Granites. Rahaman et al. (1984) believe that magma generation was initiated in the asthenosphere-lithosphere decoupling zone or below. According to these authors, thermal anomaly was due to asthenospheric plume or diapiric intrusion of peridotite along propagating fractures in the lithosphere. Equally, Sillitoe (1974) had earlier postulated that the Younger Granites was due to mantle hotspot and that the associated tin was derived directly from the mantle. On the basis of these postulations, both lower crust and upper mantle must have made significant contributions to the evolution of the Younger Granites.

## Mineralization

Mineralization in the Younger Granites has been related to different types of hydrothermal alteration processes (Kinnaird, 1979). These are:

1. Early sodic metasomatism
2. Potassic metasomatism

3. Acid metasomatism
4. Chloritization and
5. Argillization

These processes have been discussed in details by Bowben and Kinnaird (1984), Kinnaird (1985) and Kinnaird et al. (1985). Obaje (2009) gave the salient parts of their discussions as follows:

*Sodic Metasomatism*

Sodic metasomatism mostly affects the peralkaline granites and is responsible for the alteration of potash feldspar to albite, desilication and enrichment in trace and rare elements. This process is responsible for the introduction of Nb-bearing mineral (columbite) into the peraluminous biotite granites, U-bearing ores (pyrochlore) into the peralkaline granites and fergusonite into the metaluminous hornblende biotite granites.

In the peralkaline granites, other minerals that are characteristic of the sodic metasomatism are aegerine, alkaline amphiboles and Th-rich monazite. In the peraluminous granites, albitization is characterized by textural changes from medium or fine to saccharoidal fine-grained rock that is rich in either zinnwaldite or lepidolite (Bowden and Kinnaird, 1978). Other minerals that were introduced into the peraluminous rocks are cassiterite, thorite, xenotime, Th-monazite and Hf-zircon. Available evidence indicates that sodic metasomatism affects only the apical parts of the plutons (Kinnaird, 1985).

*Potassic Metasomatism*

Potassic metasomatism is characterized by the development of intermediate to ordered microcline with biotite in the compositional range annite-siderophyllite and chloritization of primary mica (Bowden and Kinnaird, 1984). Accessory monazite, zircon, cassiterite, molybdenite and occasionally wolframite are present in the phyllosilicates.

*Acid Metasomatism*

Acid metasomatism is characterized by breakdown of granitic minerals to produce new mineral assemblage. It results into the formation of sericite-topaz-quartz assemblage. In addition, microcline is transformed into micaceous aggregate, chlorite or kaolinite in rare cases. Acid metasomatism may be disseminated or form pervasive pockets or along fissures. Other accessory minerals of this alteration process are oxides and usually concentrated in the mica clusters.

*Chloritization*

Chloritic alteration is characterized by the chloritization of annitic mica, alteration of perthite to micaceous aggregate, introduction of iron and reduction in silica. It is greatest in the contact zones with the Pan-African Granites. Minerals that are associated with this process are sphalerite, chalcopyrite and fluorite.

*Argillization*

Argillic alteration is a late stage process and characterized by the conversion of feldspars to montmorillonite and kaolinite. Argillization is very limited in extent. It is an important process in areas that initially have been affected by extensive development of albite from sodic metasomatism. The clays are found in vugs in major veins, coat crystals or fill intracrystal voids in smaller veins.

## *Styles of Mineralization*

The hydrothermal processes mainly affect the biotite granites. The processes generally accompanied with the development of Sn, W, Zn, Nb, Cu, Fe, Bi, U and REE in and around the roof and marginal zones of the medium- or fine-grained granite cupolas, with veins extending up to about 2 Km into the country rocks. Obaje (2009) listed 9 different styles of mineralization as follows:

1. pegmatite pods with topaz and beryl in addition to quartz and feldspar;
2. disseminated columbite with pyrochlore ± cassiterite;
3. prejoint and postjoint pegmatites with uraninite, thorite and columbite;
4. quartz rafts, stockworks, sheeted veins and altered wall rock with cassiterite, wolframite and sulphides;
5. fissure-filling veins or loads with cassiterite, wolframite and sulphides;
6. irregular shaped replacement bodies with cassiterite and sulphides;
7. quartz veins with wolframite or scheelite, Bi minerals, cassiterite/or sulphides
8. mineralized ring dykes with cassiterite and sulphides; and
9. alluvial and eluvial deposits of cassiterite, columbite, zircon etc.

# SUMMARY OF THE GRANITE OCCURRENCE IN NIGERIA

In Nigeria, granites were emplaced at two different geological ages and tectonic environments. The older ones were emplaced during the Pan-African orogeny in the Neoproterozoic (ca. 600 Ma) and are so described as the Older or Pan-African Granites. The younger was associated with the Mesozoic anorogenic mid-plate magmatism localized to the northcentral Nigeria. These are referred to as the Younger Granites.

The Pan-African Granites represent a major petrological unit of the Basement Complex of Nigeria. The granites range in composition from tonalites through granodiorites to true granites with associated syenite and pegmatites. The Pan-African Granites are not known to contain any mineralization.

The Younger Granites occur as groups of ring dykes called ring complexes and constitute parts of a larger province of mid plate anorogenic alkaline magmatism of the Mesozoic. They Younger Granites occur in a restricted zone in the northcentral Nigeria and extends upward into the Republic of Niger. The ages of the complexes decrease southwards, indicating migratory centers of magmatic activity. The Younger Granites are essentially high-level intrusives and are mostly peralkaline with subordinate peraluminous and metaluminous suites. Over 50 ring complexes have been identified. Four major petrographic groups of the Younger Granites are distinguished: hornblende-pyroxene-fayalite granites, hornblende-biotite-granites, biotite granites and riebeckite granites with associated syenite, gabbros, rhyolites, basalts, agglomerates, tuffs and dolerites. Three evolutionary stages: early volcanic, ring dyke and caldera, and intrusive stages have been proposed for the evolution of the Younger Granites.

The albitized granites, and minor pegmatites and quartz veins are known to contain Nb, Sn, U, Li, $PO_4$, Zn, Be, Pb, Cu, W, and Th mineralization. The alluvial and eluvial deposits of Sn and Nb derived from these mineralized Younger Granites have been worked extensively in the past.

Conclusively, though general mapping of the Pan-African Granites has been done, a lot of information is still needed. This includes detailed structural mapping of each pluton to ascertain the mechanism of emplacement, the identification of the fertile granites (those granites that are parental to mineralized pegmatites) and the reasons for general lack of mineralization. The identification of the fertile plutons will aid in exploration for mineralized pegmatites that may still be unexposed at the present erosional level or yet to be discovered. In the case of the Younger Granites, much work is still needed in areas of structural and isotopic studies to be able to fully understand processes of magma generation and mechanisms of emplacement of individual ring complexes.

## REFERENCES

Affaton, P., Rahaman, M.A., Trompetts, R. and Sougy, J. (1991) The Dahomeyide orogen: tectonothermal evolution and relationships with the Volta Basin. In: Dallmeyer, R.D. and Lecorche, J.F. (eds.), The West African Orogens and Circum Atlantic Correlatives. IUGS/UNESCO project 233, *Springer Verlag*, pp. 107-122.

Ajakaiye, D.E. (1983) Deep structures of alkaline ring complexes from geophysical data. In:Abstract, international conference on alkaline ring complexes in Africa. Zaria, *Nigeria*

Ajibade, A.C. (1976) Provisional classification and correlation of schist belts of northwestern Nigeria. In: Kogbe, C.A. (ed.), *Geology of Nigeria*. Elizabethan Pub. Co. Lagos, Nigeria. pp. 85 – 90.

Ajibade, A.C. (1980) Geotectonic evolution of the Zungeru region, *Nigeria*. Unpub. Ph.D. Thesis, University College, Wales, Aberystwyth, 308 pp.

Ajibade, A.C. and Fitches, W.R. (1988) The Nigerian PreCambrian and the Pan-African orogeny. In Oluyide, P.O., Mbonu, W.C., Ogezi, A.E., Egbuniwe, I.G., Ajibade, A.C. and Umeji, A.C. (eds.), Precambrian Geology of Nigeria. *Geol. Surv. Nigeria spec. pub.*, pp. 45-53.

AJibade, A.C., Woakes, M. and Rahaman, M.A. (1987) Proterozoic crustal development in the Pan-African regime of Nigeria. In: Kroner, A. (ed.), Proterozoic Lithospheric Evolution. *Am. Geophy. Union Spec. Pub*. pp. 260-270.

Beka, F.T. and Ukaegbu, V.U. (2008) Petrology and major element geochemistry of high-K peraluminous granites in southeast Obudu plateau, southeastern Nigeria. *J. Min. Geol.* 44(2), 107-119.

Bowden, P. (1985) The geochemistry and mineralization of alkaline ring complexes in Africa. *J. Afr. Earth Sci*. 3, 17–39.

Bowden, P. and Turner, D.C. (1974) Peralkaline and associated ring complexes in the Niger-Nigeria province, West Africa. In: Sorensen, H (ed.), *The Alkaline Rocks*. Wiley, London. pp. 330-351.

Bowden, P. and Bennett, J.N. (1977) Implications of the isotopic evidence and volcanic rock types for the origin and evolution of the Nigeria Younger Granite Province, Nigeria. *J. Min. Geol.* 14(2), P58.

Bowden, P. and Kinnaird, J.A. (1978) Younger granites of Nigeria: A zinc-rich tin province. *Trans. Inst. Min. Metall.* B 78, 66–69.

Bowden, P. and Kinnaird, J.A. (1984) Geology and mineralization of the Nigerian anorogenic ring complexes. *Geologisches Jahrb (Hannover).* B56, 3–65.

Bowden, P., Van Breemen, O., Hutchison, J. and Turner, D.C. (1976) Palaeozoic and Mesozoic age trends for some ring complexes in Niger and Nigeria. *Nature.* 259, 297–299.

Burke, K.C. and Dewey, J.F. (1972) Orogeny in Africa. In: Dessauvagie, T.F.J. and Whiteman, A.J. (eds.), African Geology. University of Ibadan press, Ibadan, *Nigeria.* pp. 583–608.

Burke, K.C. and Dewey, J.F. (1973) An outline of the Precambrian plate development. In:Tarling, D.M. and Runlovn, S.K. (eds.), Implication of Continental Drift to Earth Sciences. *Academic Press*, London. pp. 1035–1045.

Caby, R. (1989) Precambrian terranes of Benin-Nigeria and northeast Brazil and late Proterozoic south Atlantic fit. *Geol. Soc. Am. Spec. paper.*230, 145-158.

Caby, R., Bertrand, J.M.L. and Black, R. (1981) Pan-African ocean closure and continental collision in the Hoggar-Iforas segment, central Sahara. In: Kroner, A.(ed.), Precambrian Plate Tectonics. Elsevier, Amsterdam. pp. 407-434.

Clemens, J.D. (1998) Observations on the origins and ascent mechanisms of granitic magmas. *J. Geol. Soc.* 155(5), 843-851.

Dada, S.S. (2006) Proterozoic evolution of Nigeria. In: Oshin, O. (ed.), The Basement Complex of Nigeria and its mineral resources (A Tribute to Prof. M. A. O. Rahaman). Akin Jinad and Co. Ibadan, *Nigeria*, pp. 29–44.

Dada, S.S., Lancellot, J.R. and Briqueu, L. (1989) Age and origin of annular charnockitic complex at Toro, northern Nigeria: U-Pb and Rb-Sr evidence. *J. Afr. Earth Sci.* 9, 227-234.

de Swardt, A.M. (1953) The geology of the country around Ilesa, Nigeria. *Geol. Surv.Bull.* No. 23, 54pp.

Ekwueme, B.N. (1993) An Easy Approach to Igneous Petrology. Univ. Calabar Press, *Nigeria.* 217pp.

Falconer, J.D. (1911) The geology and geography of Northern Nigeria. Macmillan London, 135pp.

Ferre, E., Gleizes, G. and Bouchez, J.L. (1995) Internal fabric and strike-slip emplacement of the Pan-African granite of Solli Hills, northern Nigeria. *Tectonics.* 14(5), 1205-1219.

Frost, B.R., Barnes, C.G., Collins, W.J., Arculus, R.J., Ellis, D.J. and Frost, C.D. (2001) A Geochemical classification for granitic rocks. *J. of Petr.* 42, (11) 2033 2048.

Jones, H.A. and Hockey, R.D. (1964) The geology of part of southwestern Nigeria. *Geol. Surv. Bull.* No 31, 87pp.

King, B.C. and de Swardt, A.M. (1949) The geology of Osi area, Ilorin Province, Nigeria. *Geol. Surv. Bull.* No. 20, 90pp.

Kinnaird, J.A. (1979) Mineralization associated with the Nigerian Mesozoic ring complexes: Studies in geology. *Salamanca.* 14, 189–220.

Kinnaird, J.A. (1985) Hydrothermal alteration and mineralization of the alkaline anorogenic ring complexes of Nigeria. *J. Afr. Earth Sci.* 3, 229–251.

Kinnaird, J. A., Bowden, P., Ixer, R.A. and Odling, N.W.A. (1985) Mineralogy, geochemistry and mineralization of the Ririwai complex, northern Nigeria. *J. Afr. Earth Sci.* 3, 185–222.

Matheis, G. and Caen-Vachette, M. (1983) Rb-Sr isotopic study of rare-metal-bearing and barren pegmatites in the Pan-African reactivation zone of Nigeria. *J. Afr. Earth Sci.* 1, 35–40.

Mc Curry, P. (1976) The geology of the Precambrian to Lower Paleozoic rocks of Northern Nigeria:- A review. In: Kogbe, C.A. (ed.), *Geology of Nigeria*. Elizabethan Pub. Co., Lagos, Nigeria. pp. 15-39.

McCurry, P. and Wright, J.B. (1977) Geochemistry of calc-alkaline volcanics in northwestern Nigeria and possible Pan-African suture zone. *Earth Planet Sci Letters.* 37, 90-96.

Mullan, H.S. (1979) Structural distinction between a metasedimentary cover and an underlying basement in the 600 Ma old Pan-African domain of northwestern Nigeria. Discussion. *Geol. Soc. Amer. Bull.* 90, 983-984.

Obaje, N.G. (2009) Geology and mineral resources of Nigeria. Lecture notes in Earth Sciences 120, Springer Verlag, 219 pp.

Odeyemi, I.B. (1977) On the petrology of the basement rocks around Igarra, Nigeria. Unpub. Ph.D. Thesis, University of Ibadan, *Nigeria*. 233pp.

Odigi, M.I. (1986) Accessory minerals in the Oban Massif granitic plutons of southeastern Nigeria- their qualitative and quantitative significance in fertility studies. *J. Afr. Earth Sci.,* 5(2), 163-166.

Okonkwo, C.T. and Winchester, J.A. (2004) Geochemistry of granitic rocks in Jebba area, southwestern Nigeria. *J. Min. Geol.* 40(2), 95-100.

Olarewaju, V.O. (1981) Geochemistry of the charnockitic and granitic rocks of the Basement Complex around Ado-Ekiti-Akure, SW, Nigeria. Unpublished Ph.D.Thesis, University of London, 383pp.

Pearce, J.A., Harris, N.B. and Tindle, A.G. (1984) Trace element discrimination diagrams for the tectonic interpretation of granitic rocks. *J. of Petr.* 25, 956-983.

Petters, S.S. (1991) Regional geology of Africa: Lecture notes in Earth Sciences 40, Springer Verlag, 722 pp.

Rahaman, M.A. (1973) The geology of the district around Iseyin, western Nigeria. Unpub. Ph.D Thesis, University of Ibadan, *Nigeria*. 268pp.

Rahaman, M.A. (1976) Review of the basement geology of Southwestern Nigeria. In: Kogbe, C.A. (ed.), *Geology of Nigeria*. Elizabethan Pub. Co., Lagos. pp. 41-58.

Rahaman, M.A. (1988) Recent advances in the study of the Basement Complex of Nigeria: In Oluyide, P.O., Mbonu, W.C., Ogezi, A.E., Egbuniwe, I.G., Ajibade, A.C. and Umeji, A.C. (eds.), Precambrian Geology of Nigeria. *Geol. Surv. Nigeria spec. pub.*, pp. 11-41.

Rahaman, M.A. and Ocan, O.O. (1978) On relationships in the PreCambrian migmatitic gneisses of Nigeria. *J. Min. Geol.* 15, 23 – 32.

Rahaman, M.A., Van Breeman, O., Bowden, P. and Bennett, J.N. (1984) Age migration of anorogenic ring complexes in Northern Nigeria. *J. Geol.* 92,173–184.

Rahaman, M.A., Tubosun, I.A. and Lancellot, J.R. (1991) U-Pb geochronology of potassium syenites from southwestern Nigeria and timing of deformation events during the orogeny. *J. Afr. Earth Sci.* 13 (3-4), 387-395.

Rahaman, M.A.O., Ajibade, A.C., Olarewaju, V.O. and Ocan, O. O. (2005) Manual for course on mapping of migmatite and granite terrains. 63 pp.

Rickwood, P.C. (1989) Boundary lines within petrologic diagrams which use oxides of major and minor elements. *Lithos*. 22, 247-263.

Sillitoe, R.H. (1974) Tin mineralization above mantle hot spots. *Nature*. 248, 497.

Trunswell, J.F. and Cope, R.N. (1963) The geology of parts Niger and Zaria provinces, northern Nigeria. *Geol. Surv. Nigeria, Bull*. No. 29.

Turner, D.C. (1976) Structure and petrology of the Younger Granite ring complex, 3-157.

Ukwang, E. and Ekwueme, B.N. (2009) Geochemistry and geotectonic study of granitic rocks of southwest Obudu plateau, southeastern Nigeria. *J. Min. Geol*. 45(1), 73-82.

Van Breemen, O., Pidgeon, R.T. and Bowden, P. (1977) Age and isotopic studies of some Pan-African granites from north-central Nigeria. *Precambrian Research*. 4, 307–319.

Woakes, M., Rahaman, M.A. and Ajibade, A.C. (1987) Some metallogenetic features of the Nigerian basement. *J. Afr. Earth Sci*. 6, 54–64.

In: Granite  
Editors: Miroslava Blasik and Bogdashka Hanika

ISBN: 978-1-62081-566-3  
© 2012 Nova Science Publishers, Inc.

*Chapter 5*

# PETROLOGY, GEOCHEMISTRY AND ORIGIN OF TOPAZ GRANITE

### *Miloš René*
*Institute of Rock Structure and Mechanics,*
*Academy of Sciences of the Czech Republic, Czech Republic*

## ABSTRACT

Topaz granites are alkali feldspar granites containing albite, quartz, K-feldspar, lithium mica and topaz, which are characterized by their an extreme enrichment in F (up to 4 wt.%), enrichment in Li, Rb, Cs, and also in Sn and W. These granites are usually subdivided into P-rich and P-poor subtypes. In the Saxothuringian Zone of the Bohemian Massif (Central European Hercynides) both types occur, which are connected with economically interested Sn-W-Nb-Ta-Li mineralization. The most important occurrences of the P-rich granite suite are stocks in the area of the Ehrenfriedersdorf and Horní Slavkov–Krásno ore districts, whereas Altenberg and Cínovec/Zinnwald granite stocks represent the P-poor granite suite.

The P-rich granites host large amounts of apatite, together with monazite and relatively scarce xenotime. In P-poor granites apatite is absent and the low whole-rock P content (< 0.1 wt.% $P_2O_5$) is reflected by variable amounts of monazite and xenotime. These granites also have a significant occurrence of thorite, Hf-highly enriched zircon with low P amount, bastnäsite, synchysite-(Y), chernovite-(Y) and complex Zr(Hf)-Th-REE phases. The P-rich granites are late-collisional, strongly peraluminous S-type granites (A/CNK = 1.1–1.5) with enrichment in P and depletion in REE and other HFSE. The distinctly high phosphorus contents of these granites (0.3–0.7 wt.% $P_2O_5$), high Al content and low Ca abundance (<0.2 wt.% CaO) facilitated the incorporation of P into the structure of late crystallized silicate minerals (albite, K-feldspar, topaz, zircon). The P-poor granites are post-collisional, slightly peraluminous granites (A/CNK = 1.1–1.2), with very low P and high REE and HFSE. These granites display some signatures of A-type granites. Contrasting patterns in the behavior of phosphorus in P-rich and P-poor granites could be explained by the strong control of peralumosity on apatite solubility in the granite melt. The significant enrichment in P of P-rich granites leads to the formation of P- and Al-bearing complexes in residual melts resulting in high P-contents of feldspars, topaz and zircon. The P-rich topaz granites were derived by anatectic melting of upper crustal, predominantly metasedimentary sequences of the Saxothuringian Zone, whereas the P-poor topaz granites were derived by small-volume partial melting of lower crustal felsic granulites and/or metagranites.

## INTRODUCTION

The fluorine enriched granitic melts at the origin of topaz granites particularity reflect the composition of their source rocks. The topaz granites are characterized by a wide variety of lithophile elements (e.g., Li, Rb, Cs) as well as selected ore metals (Sn, W, Nb, Ta). They can be subdivided according to Taylor [1] into a low-$P_2O_5$ (< 0.4 wt.% $P_2O_5$) and high-$P_2O_5$ (> 0.4 wt.% $P_2O_5$) subtype. From a geotectonic point of view the P-poor granites could be classified as A-type granites, whereas the P-rich granites have signatures of highly evolved S-type granites [2]. The Saxothuringian Zone of the Central European Hercynides is one of the most significant geotectonic and metalogenic provinces of world with a huge occurrence of both topaz granite types with narrow relations to the Sn-W-Nb-Ta-Li mineralization. The aim of this paper is to demonstrate selected mineralogical and geochemical features to recognize the dichotomy of origin and sources for these contrasting varieties of topaz granites.

## GEOLOGICAL SETTING

Topaz granites form two distinct magmatic suites in the Saxothuringian Zone of the Bohemian Massif [3, 4, 5, 6] (Figure 1). The first suite represents high-evolved P-rich S-type granites. These granites occur in granite stocks of the western and central parts of the Krušné Hory/Erzgebirge area. From a geotectonic point of view these granites are part of the Hercynian late-collisional distinctly fractioned magmatic suite [4, 5]. The most important occurrences of the P-rich suite are granite stocks in the area of the Ehrenfriedersdorf Sn-W ore district (Germany) [7] and in the Horní Slavkov-Krásno ore district (Czech Republic) [3, 8]. Other, relatively small occurrences of P-rich granites are the Podlesí topaz-albite granite stock [3, 9] and the Geyersberg granite stock [10].

The Ehrenfriedersdorf Sn-W ore district is situated in the northwestern flank of the Central Erzgebirge Anticlinal Zone. In this area are evolved some small, mostly hidden topaz granite stocks (Vierung, Sauberg, Geyersberg, Ziegelberg and Greifenstein) which are part of a bigger Annaberg pluton. The topaz-bearing granites in the stocks are subdivided into four intrusive phases (A–D) [11]. The inner structure of these stocks is only partly stratified. The fine- to medium-grained topaz-bearing monzogranite (phase C) represents a main intrusive phase in these stocks. The youngest, partly altered topaz-albite granite (phase D) typically occurs as dykes in older magmatic phases of these granite stocks. The Horní Slavkov-Krásno ore district comprises topaz-bearing granite stocks evolved along the southeastern margin of the Krudum granite body in the Slavkovský les Mts. area [8, 12, 13, 14] (Figure 1). The inner structure of these stocks (Hub, Schnöd and Vysoký Kámen) is well stratified, comprising partly greisenized topaz-albite granites, leucocratic topaz-albite granites and layers of alkali-feldspar syenites. The Podlesí granite stock forms a small stock on the northern edge of the Blatná granite body in the western part of the Krušné Hory/Erzgebirge area. This stock consists predominantly of topaz-albite granite, which can be divided into two sub-facies. The uppermost part of the stock is intruded with generally flat lying dykes of topaz-bearing dyke granite. The Krudum and Blatná granite bodies are subsidiary intrusions of the Karlovy Vary–Eibenstock pluton, which represent main granite intrusion in western part of the Krušné Hory/Erzgebirge area. The second topaz granite suite is formed by P-poor, A-type granites. These granites include small bodies and stocks with important Sn-W ore mineralization of the volcanic-plutonic complex of the eastern Krušné Hory/Erzgebirge area (Altenberg, Zinnwald/Cínovec, Sadisdorf, Sachsenhöhe, Krupka). This magmatic suite is part of Hercynian post-collisional granite magmatism in this area. The suite represents the final magmatic activities within the Central European Hercynian granite belt [5, 15]. The Cínovec granite stock is a relatively small, elliptical, vertical stratified body in the central part of the Altenberg-Teplice caldera (Figure 2). The borehole CS-1, located in the center of granite stock, transected lepidolite-bearing granite at the top of the section (about 90 m thick), an intermediate zone of zinnwaldite-bearing granite (thickness about 640 m) and a lower zone of protolithionite-bearing granite to the depth 1596 m [17, 18]. Small topaz granite stocks of similar composition were recently found in the central part of the Krušné Hory/Erzgebirge area near Seiffen [19] and Hora Svaté Kateřiny [20].

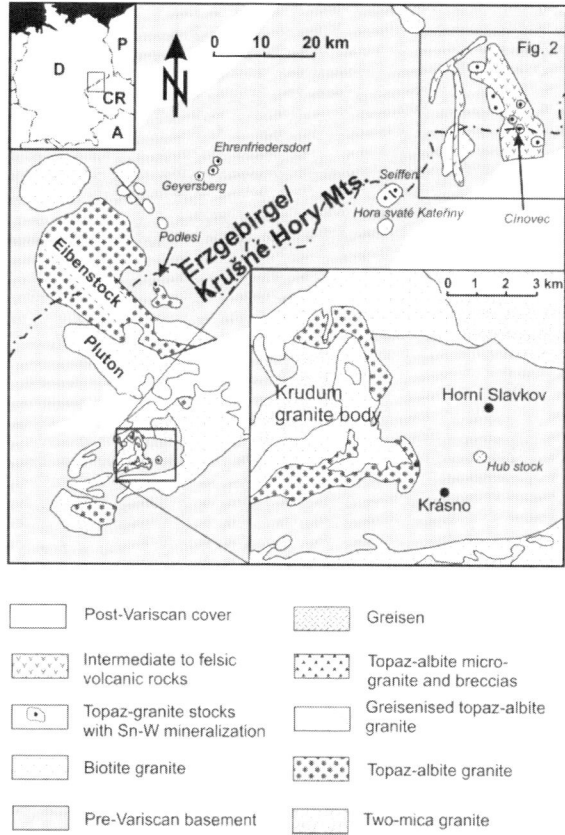

Figure 1. Distribution of late-Hercynian igneous rocks in the Krušné Hory/Erzgebirge area, with the Krudum granite body and associated granite stocks shown as inset map (modified from Müller et [14]).

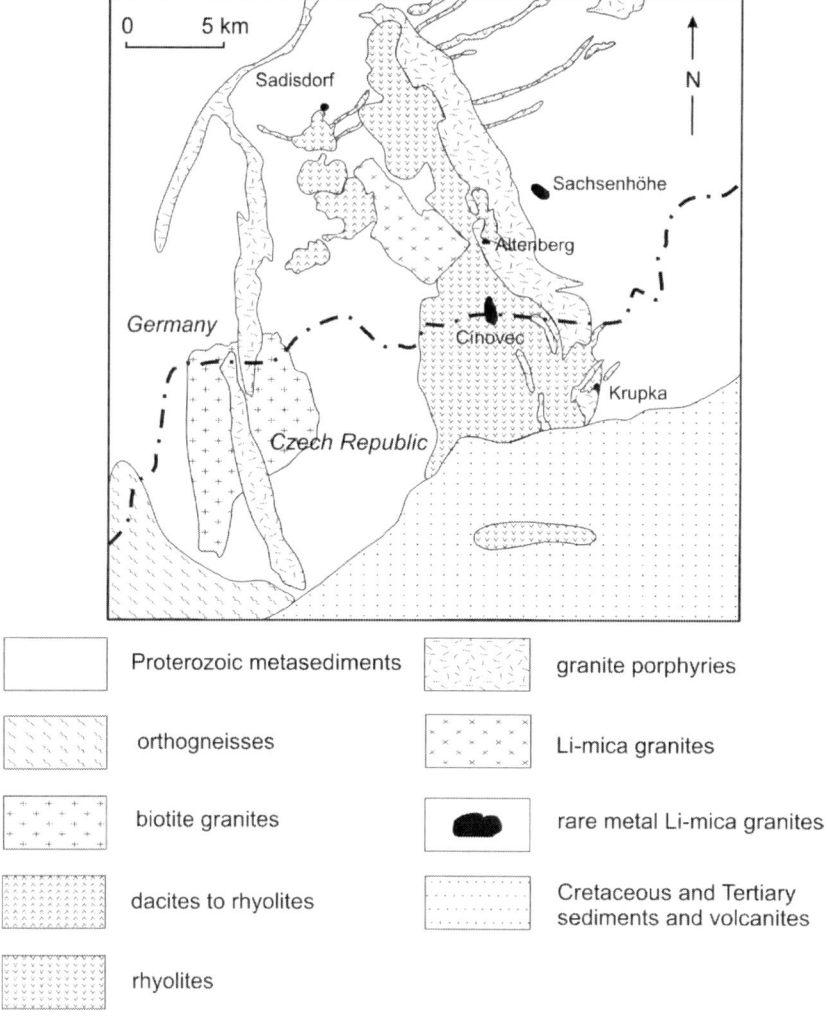

Figure 2. Geological map of the volcano-plutonic complex of the eastern Krušné Hory/Erzgebirge with location of the Cínovec granite stock (after Seifert and Kempe [16]).

## ANALYTICAL METHODS

The whole-rock composition of the topaz granites was determined for 36 representative samples from the Krásno–Horní Slavkov ore district, the Podlesí, Geyersberg, Sauberg and Cínovec granite stocks (Table 1). Major elements were determined by X-ray fluorescence spectrometry using the PANalytical Axios Advanced spectrometer at Activation Laboratories Ltd., Lancaster, Canada.

The FeO content was measured by titration, whereas the loss on ignition (LOI) was determined gravimetrically at the Analytical Laboratory of the Institute of Rock Structure and Mechanics (Academy of Sciences of the Czech Republic), Prague, Czech Republic. The F content was determined by using an ion-selective electrode in the same laboratory. Trace elements were determined by inductively-coupled-plasma mass-spectrometry (ICP MS) using a Perkin Elmer Sciex ELAN 6100 ICP mass spectrometer at Activation Laboratories Ltd.

Table 1. Representative chemical analyses of topaz granites from the Krušné Hory/Erzgebirge area

| wt. % | 1007 | 1283 | 1285 | 1366 | 630 | 631 | 633 | 634 |
|---|---|---|---|---|---|---|---|---|
| | Krásno | Krásno | Krásno | Krásno | Cínovec | Cínovec | Cínovec | Cínovec |
| $SiO_2$ | 73.28 | 73.71 | 74.34 | 74.72 | 76.04 | 76.79 | 76.31 | 77.13 |
| $TiO_2$ | 0.06 | 0.04 | 0.05 | 0.05 | 0.02 | 0.03 | 0.06 | 0.05 |
| $Al_2O_3$ | 14.81 | 15.40 | 15.05 | 15.01 | 13.24 | 12.65 | 12.27 | 12.31 |
| $Fe_2O_3$ | 0.04 | 0.01 | 0.06 | 0.00 | 0.40 | 0.47 | 0.67 | 0.51 |
| FeO | 0.91 | 0.77 | 0.81 | 0.75 | 0.47 | 0.44 | 0.48 | 0.55 |
| MnO | 0.08 | 0.04 | 0.05 | 0.05 | 0.05 | 0.04 | 0.03 | 0.03 |
| MgO | 0.17 | 0.05 | 0.05 | 0.05 | 0.05 | 0.07 | 0.11 | 0.07 |
| CaO | 0.60 | 0.40 | 0.38 | 0.37 | 0.79 | 0.44 | 0.64 | 0.45 |
| $Na_2O$ | 3.25 | 3.72 | 3.49 | 3.50 | 3.27 | 3.42 | 3.32 | 3.30 |
| $K_2O$ | 4.75 | 4.12 | 3.69 | 4.06 | 4.69 | 4.79 | 5.01 | 5.16 |
| $P_2O_5$ | 0.31 | 0.27 | 0.27 | 0.30 | 0.02 | 0.02 | 0.03 | 0.02 |
| $H_2O^+$ | 1.20 | 0.05 | 0.41 | 0.19 | 0.51 | 0.51 | 0.51 | 0.42 |
| $H_2O^-$ | 0.00 | 0.00 | 0.32 | 0.14 | 0.17 | 0.21 | 0.22 | 0.11 |
| F | 0.68 | 0.12 | 0.16 | 0.16 | 0.35 | 0.70 | 0.45 | 0.40 |
| O=F | -0.29 | -0.05 | -0.07 | -0.07 | -0.15 | -0.29 | -0.19 | -0.17 |
| Total | 99.85 | 98.65 | 99.06 | 99.28 | 99.92 | 100.29 | 99.92 | 100.34 |
| A/CNK | 1.28 | 1.36 | 1.44 | 1.39 | 1.07 | 1.09 | 1.06 | 1.04 |
| ppm | | | | | | | | |
| Ba | 26 | 23 | 21 | 17 | 64 | 63 | 53 | 31 |
| Rb | 1030 | 1150 | 1160 | 1031 | 1550 | 1250 | 682 | 827 |
| Sr | 27 | 14 | 13 | 10 | 7 | 8 | 14 | 8 |
| Li | 762 | 729 | 735 | 704 | 516 | 416 | 159 | 195 |
| Y | 7 | 10 | 10 | 9 | 122 | 78 | 142 | 149 |
| Zr | 37 | 53 | 40 | 38 | 92 | 98 | 146 | 82 |
| U | 10 | 27 | 22 | 9 | 51 | 26 | 27 | 40 |
| Th | 6 | 9 | 7 | 6 | 34 | 32 | 55 | 32 |

Since the analytical procedure for ICP MS involves lithium metaborate/tetraborate flux fusion, the Li concentration was analyzed separately by atomic absorption spectrometry on a Varian 220 spectrometer at the Institute of Rock Structure and Mechanics.

For presentation of geochemical features of both topaz granite suites additional analyses from the Podlesí, Seiffen, Hora Svaté Kateřiny and Cínovec stocks published by Breiter [6, 20, 21], Förster and Rhede [19] were used.

## MINERALOGICAL AND GEOCHEMICAL COMPOSITION

### Horní Slavkov-Krásno Ore District

The partly greisenised topaz-albite alkali feldspar granite (TAG) is medium-grained, equigranular rock which consists of quartz, albite ($An_{0-2}$), potassium feldspar, lithium mica (protolithionite) and topaz. Fluorapatite, zircon, Nb-Ta-Ti oxides, xenotime-(Y), monazite-(Ce), uraninite and coffinite are common accessory minerals.

TAG is a peraluminous rock with aluminous saturation index (A/CNK) ranging from 1.1 to1.5 (Figure 3a). Compared to common Ca-poor granites [22], it is enriched in incompatible elements such as Li (160–820 ppm), Rb (830–1500 ppm), Cs (38–150 ppm), Sn (19–6200 ppm), Nb (18–83 ppm), Ta (8–53 ppm), W (4–62 ppm) and poor in Mg (0.1–0.2 wt.% MgO), Ca (0.3–1.0 wt.% CaO), Sr (12–50 ppm), Ba (21–81 ppm) and Zr (20–55 ppm) (Figure 3b, c). The granite is distinctly enriched in phosphorus (0.3–0.4 wt.% $P_2O_5$) and fluorine (0.1–0.8 wt.% F).

Granite is distinctly depleted in REE ($\Sigma REE$ = 12–46 ppm) (Figure 3d). A high degree of magmatic fractionation is reflected in the low K/Rb ratio (15–47).

### Podlesí Stock

The uppermost sub facies of stock granite at the Podlesí stock is represented by fine-grained, porphyric topaz-albite alkali feldspar granite. The lower facies constitutes the main part of this stock.

This topaz-albite alkali feldspar granite is medium-grained (0.2–0.8 mm), equigranular rock, which consists of albite, potassium feldspar, quartz, Li-mica (protolithionite) and topaz. Among accessories, apatite prevails over monazite, zircon, cassiterite, Nb-Ta-Ti oxides, wolframite, ilmenite, and hematite.

This granite is highly peraluminous rock (A/CKN = 1.17–1.39), enriched in Rb (985–1840 ppm), Cs (68–197 ppm), Nb (28–72 ppm), Ta (8–39 ppm) and poor in Mg (0.1–0.3 wt.% MgO), Ca (0.2–0.8 wt.% CaO), Ba (3–22 ppm), Sr (7–72 ppm), Zr (15–65 ppm) and Th (3–8 ppm).

The granite is also distinctly enriched in P (0.3–0.7 wt.% $P_2O_5$) and fluorine (0.9–1.8 wt.% F). Its bulk content of REE is low ($\Sigma REE$ = 7–26 ppm). A high degree of magmatic fractionation is reflected in the low K/Rb ratio (20–40).

## Sauberg Stock

The magmatic phase of group C, which represents the main granite phase of the Sauberg stock at the Ehrenfriedersdorf ore deposit is fine- to medium-grained, equigranular rock, consisting of quartz, albite ($An_{5-10}$), potassium feldspar, Li-mica and topaz. Accessory minerals are represented by apatite, zircon and rarely Nb-Ta-bearing rutile, columbite-(Fe), cassiterite and W-ixiolite. The granite is peraluminous rock (A/CKN = 1.2–1.4), enriched in Rb (1210–1900 ppm), Cs (95–126 ppm), Li (380–1570 ppm) and Nb (22–59 ppm). This granite is poor in Mg (0.05–0.2 wt.% MgO), Ca (0.3–0.7 wt.% CaO), Ba (10–38 ppm), Zr (26–49 ppm), Y (4–14 ppm) and Th (4–10 ppm). Its bulk amount of REE is also low ($\Sigma$REE = 6–16 ppm). The low K/Rb ratio (15–34) of this granite reflects relatively high magmatic fractionation.

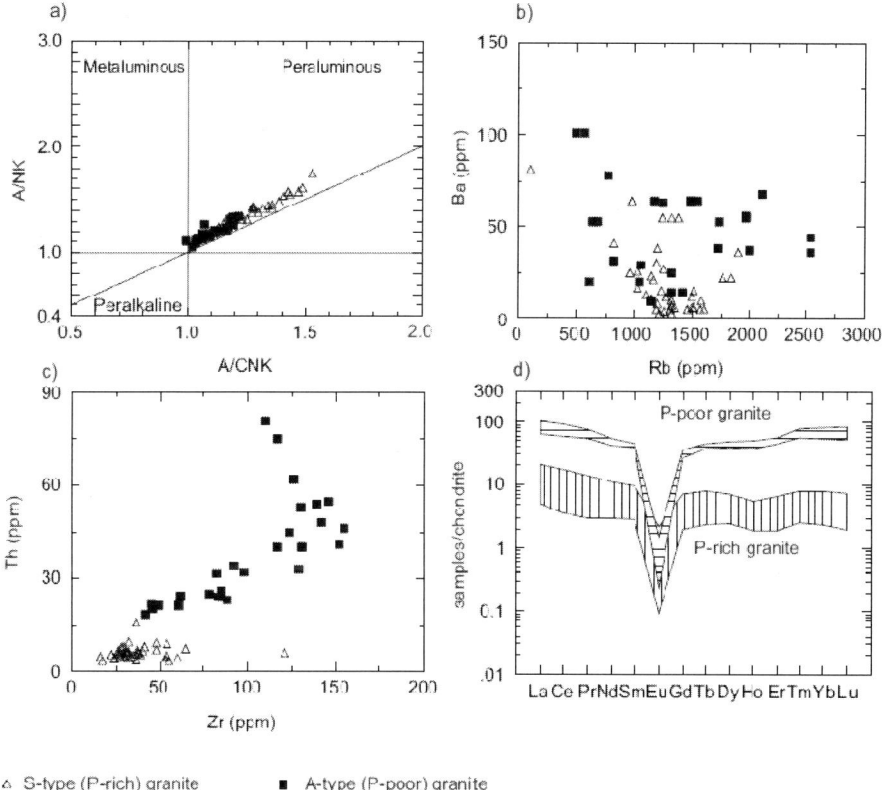

Figure 3. Geochemical features of topaz-bearing granites from the Krušné Hory/Erzgebirge area. a) Shand´s diagram after Maniar and Piccoli [55] b) Plot of Rb vs. Ba, c) Plot of Th vs. Zr, d) Chondrite normalized REE patterns. Normalizing values are from Taylor and McLennan [56].

## Geyersberg Stock

In the relatively small granite stock near Geyer town in the Ehrenfriedersdorf ore district the middle- to fine topaz-albite alkali feldspar granites of group D were investigated. These granites consist of albite, potassium feldspar, quartz, Li-mica and topaz.

The apatite, Nb-Ta-bearing rutile, zircon and W-ixiolite are common accessories. The granite is peraluminous (A/CKN = 1.2–1.4), enriched in P (0.5–0.8 wt.% $P_2O_5$), Rb (1190–1660 ppm), Cs (28–42 ppm), Nb (33–101 ppm), Ta (37–76 ppm) and depleted in MgO (0.04–0.05 wt.% MgO), Ca (0.5–0.6 wt.% CaO), Zr (14–17 ppm), Th (6–13 ppm) and Y (0.5–1 ppm). The analysed granites show extremely low bulk concentrations of REE ($\Sigma REE$ = 0.5–0.8 ppm). Their K/Rb ratio (21–27) is also low. The low concentrations of Zr, Th, Y and REE together with a low K/Rb ratio reflect high magmatic fractionation of examined granites from the Geyersberg granite stock.

## Cínovec Granite Cupola

The zinnwaldite-bearing granite is predominantly a medium grained rock containing quartz, albite ($An_{0-5}$), potassium feldspar and zinnwaldite. The topaz, fluorite, Nb-Ta-Ti-oxides, cassiterite, zircon, Th-rich monazite-(Ce), bastnäsite, REE oxyfluorides and hydroxyfluorides, pyrochlore, complex Zr-(Hf)-Th-REE phases, wolframite and scheelite represent the accessory mineral assemblage. The granite is a slightly peraluminous rock with A/CNK ranging from 1.1 to 1.2 (Figure 3a). Compared to typical A-type granites [23], it is enriched in Rb (584–2371 ppm), Nb (56–86 ppm), Ga (27–58 ppm) and depleted in Zr (24–98 ppm), Y (42–122 ppm) and Ce (81–87 ppm) (Figure 3b, c). Granite is distinctly enriched in REE ($\Sigma REE$ = 226–251 ppm) (Figure 3d). A high degree of magmatic fractionation is reflected in the low K/Rb ratio [14–32].

Slightly porphyritic protolithionite-bearing syenogranite contains quartz, albite ($An_{5-10}$), potassium feldspar and protolithionite. Topaz, zircon, xenotime-(Y), thorite, synchysite-(Y), synchysite-(Ce), Th-poor monazite-(Ce) and rutile are accessories. This granite is also slightly peraluminous with A/CNK ranging from 1.0 to 1.1 (Figure 3a). Compared with zinnwaldite-bearing granite, it is depleted in Rb (615–830 ppm), Nb (41–57 ppm) and Ga (25–26 ppm), but it is enriched in Zr (82–146 ppm), Y (116–149 ppm) and Th (32–75 ppm) (Figure 3b, c). Granite is also distinctly enriched in REE ($\Sigma REE$ = 190–245 ppm) (Figure 3 d). Its lower degree of magmatic fractionation is reflected in a higher K/Rb ratio [52–66].

## GEOTECTONIC CLASSIFICATION OF TOPAZ GRANITES

From a geotectonic point of view the topaz granites could be classified as S-type, I-type and A-type granites. The highly popular geotectonic classification according to granite melts source on S- and I-type granites was first proposed by Chappell and White [24]. In time with this concept the S-type granites represent sedimentary protolith, while the I-type granites represent igneous protolith. A large number of subsequent studies have reinforced the validity of the geotectonic distinction between I- and S-type granites (e.g., [2, 25]). However, for distinguishing various groups of high-evolved granites with relatively high contents of lithophile elements (Li, Rb, Cs) and fluorine, this classification is partly unusable. According to these considerations some other classifications of fractionated felsic granites were proposed. To distinguish middle alkaline granites that occur in rift zones and stable continental blocks from other fractionated felsic granites and unfractionated I- and S-type granites the concept of A-type granites was later proposed [26]. To distinguish this new granite type the following geochemical features were proposed: high $FeO_{tot.}/(FeO_{tot.} + MgO)$ ratios, high abundances of high field strength elements (HFSE) and trivalent REE, together with low abundance of Fe-Mg trace elements (Cr, V, Ni, Cu), Ba, Sr and Eu. In this new granite group were included peralkaline, metaaluminous and weakly peraluminous granites in the anorogenic geological setting. Later, for distinguishing I-, S- and A-type granite groups, various geochemical discriminating rules and diagrams evolved [23, 27, 28, 29]. Collins et al. [27] proposed for recognition of A-type granites from I- and S-type granites Ga/Al ratio. The discrimination feature was based on preferential excluding of Ga from the anorthite structure, relative to Al [30] and on high stability of $GaF_6^{3-}$ complex at high temperatures in comparison with $AlF_6^{3-}$ complex [31]. Thus, according Collins et al. [27] production of the A-type granite melt Ga would partition into melt more readily than Al, and would also be preferentially retained in that melt during crystallization. The P-poor granites from the Krušné Hory/Erzgebirge area display Ga/Al ratio from 3.9 to 7.0, whereas for P-rich granites from the same area a Ga/Al ratio from 2.1 to 9.7 is significant. Consequently, both examined granite groups display usually high Ga/Al ratios (Ga/Al > 2.6), which according to Collins et al. [27] could be characteristic only for A-type (P-poor) granites. The high Ga/Al ratio in both granite groups can be explained by extreme enrichment of fluorine and accordingly by occurrence of $GaF_6^{3-}$ complex in both granite melts. Whalen et al. [23] also demonstrated that highly fractionated felsic I- and S-type granites display high Ga/Al ratios which overlap those of typical A-type granites. Therefore they proposed a group of additional discrimination diagrams, which are based not only on Ga/Al ratio, but also on concentrations of Zr, Nb, Ce, Y, Zn and values of agpaitic index. In the case of topaz granites from the Krušné Hory/Erzgebirge area, the best discrimination between S- and A-type granites show concentrations of Ce and alternatively Y vs. Ga/Al ratio (Figure 4). For discrimination between fractionated I-, S-type granites and A-type granites Whalen et al. [23] also proposed a plot of Zr+Nb+Y vs. $FeO_{tot.}/MgO$ which offers better discriminiation results of both granite groups (Figure 5).

However, all above-mentioned geochemical constraints of the A-type group have been in the last twenty years. The matter of much debate and three main reasons for declination of the A-type granite group were postulated: (i) A-type classification is inconsistent with widely used I-S-type classification, (ii) the A-type granites overlap compositionally with felsic I-type granites, (iii) the tectonic definition of A-type granites as an anorogenic granite group is confusing.

For these reasons, some authors considered this granite group redundant (e.g., [32, 33]). Accordingly to extreme enrichment of topaz granites in F, Li, Rb, Cs, Nb, Ta, Sn and W, Taylor [1] proposed subdivision of these granites on low-$P_2O_5$ and high-$P_2O_5$ subtypes which gives more persuasive discrimination of both topaz granite groups occurring in the Krušné Hory/Erzgebirge area (Figure 6).

Later a new discriminating feature based on $\delta^{18}O$ concentrations was used [34]. This discriminating feature is very good and usable in the case of topaz granites from the Krušné Hory/Erzgebirge area. The distinctly higher $\delta^{18}O$ values in P-rich granites (10.3–13.4) are contrasted with lower $\delta^{18}O$ values founded in P-poor granites from this area (5.2–8.4). The higher $\delta^{18}O$ values show on significant metapelitic component in potential protolith, whereas low $\delta^{18}O$ values indicate a meta-igneous protolith (Figure 7).

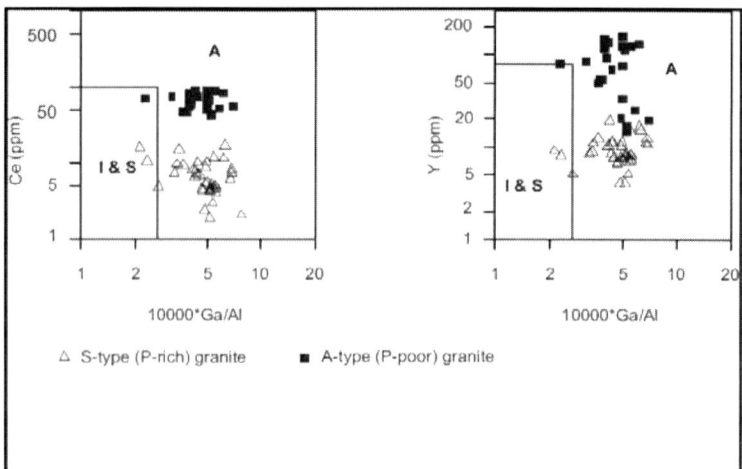

Figure 4. 10000*Ga/Al vs. Ce and Y according Whalen et al. [23] of S-type and A-type granites from the Krušné Hory/Erzgebirge area.

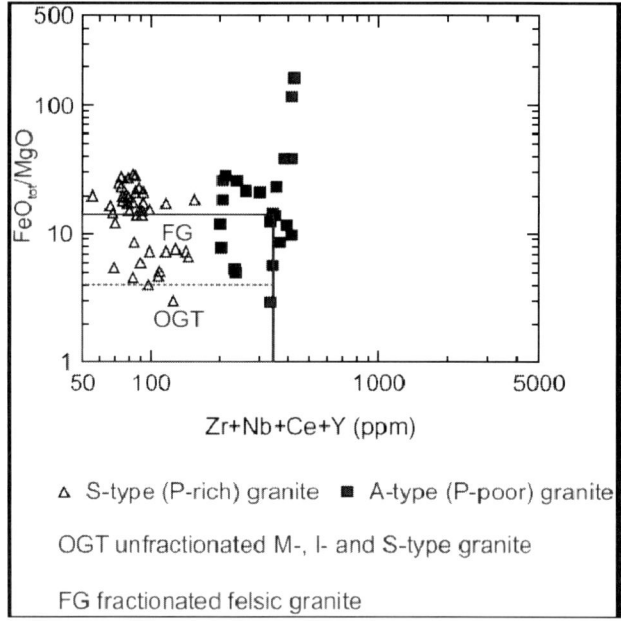

Figure 5. Zr+Nb+Ce+Y versus $FeO_{tot}$/MgO plot according Whalen et al. [23] of S-type and A-type granites from the Krušné Hory/Erzgebirge area.

Recently the A-type granites concept and principles of high-evolved granite classification matters is part of a controversial debate [2, 33]. Bonin [2], after very extensive discussion in all aspects of A-type granites, concluded that A-type granites "form a comparatively small, yet distinctive, group in Earth´s granite realm". In front with this highly positive assessment of A-type granite term, Frost and Frost [33] recently recommend the term A-type granite be replaced by the non-genetic term ferroan granites. Frost et al. [35] already introduced the term ferroan granite in their geochemical classification of granitic rock. In their last paper based on detailed analyses of A-type granites used to define this granite type by Loiselle and Wones [26] and Collins et al. [27] concluded that most granites referred to as A-type granites are ferroan granites. However, in the case of highly fractionated topaz-bearing granites from the Krušné Hory/Erzgebirge area, both granite groups are ferroan granites (Figure 8).

Consequently, this diagram cannot be applied to discrimination of both granite groups so the proposed recommendation to replace the term A-type granites by the term ferroan granite could be matter of additional discussion.

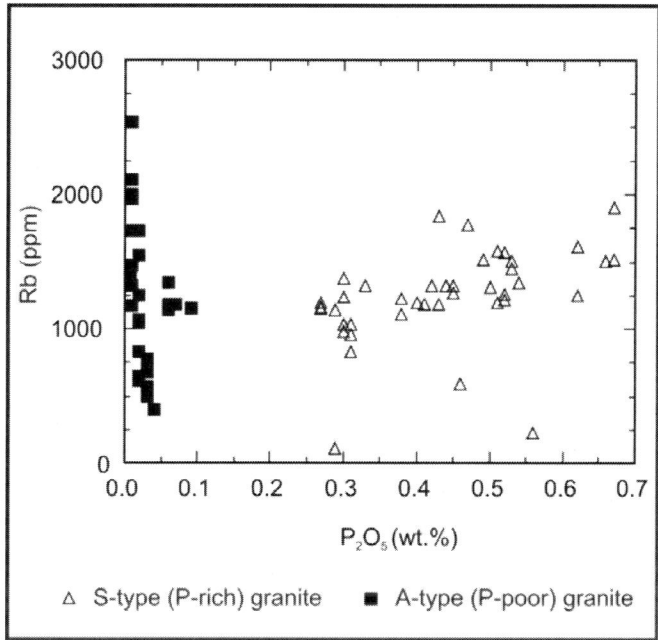

Figure 6. Rb vs. $P_2O_5$ plot of of S-type and A-type granites from the Krušné Hory/Erzgebirge area.

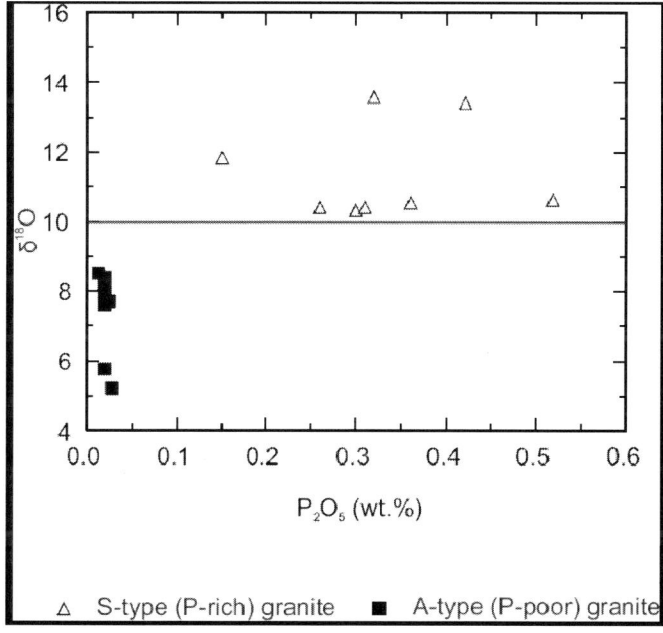

Figure 7. $\delta^{18}O$ vs. $P_2O_5$ plot of of S-type and A-type granites from the Krušné Hory/Erzgebirge area.

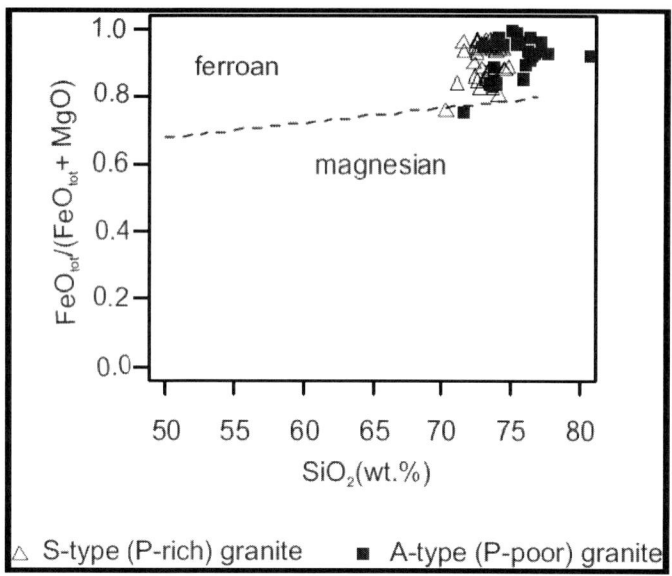

Figure 8. $FeO_{tot.}/(FeO_{tot.} + MgO)$ vs. $SiO_2$ plot according Frost and Frost [33] of S-type and A-type granites from the Krušné Hory/Erzgebirge area.

# EVIDENCE FOR DIFFERENT SOURCES

The crucial question connected to the origin of both topaz granite types is the source of these granites and the mechanism at the origin of granitic melts. The high fractionation of topaz granites reflected by high contents of lithophile elements (Li, Rb, Cs), high values of Ta/Nb ratio together with low K/Rb ratio make the identification of their source materials very difficult. In the case of topaz granites from the Krušné Hory/Erzgebirge area, their nearly contemporaneous emplacement and evolution is also significant [5, 6, 36]. The root zones of the granites are not exposed and the topaz granites are free of enclaves or restite minerals that might give some information on the source lithologies. Breiter [6] concluded that both topaz granite groups originated from a concurrent protolith formed by a mixture of fertile quartz-feldspathic and pelitic lithologies enriched in lithophile elements, tin and tungsten. According to this paper, the differences in chemical composition of emplaced granite melts originated from local changes in the proportion of quartz-feldspathic and pelitic lithologies, different PT- melting conditions, the degree of the melting, and the degree of metamorphic dehydration of the protolith prior to the melting. However, Breiter [6] ignored expressive differences between both granite groups in the REE concentrations (distinctly high enrichment of all REE in P-poor granites) and differences in $\delta^{18}O$ concentrations [17, 34]. The initial $\varepsilon_{Nd}$ values (−4.5 to −7.4) of P-rich topaz granites correspond to those of the two-mica S-type granites of neighboring areas of the western part of the Bohemian Massif [4]. However, in P-poor topaz granites, the initial $\varepsilon_{Nd}$ values vary between −1.8 and −4.4 [20, 36], which correspond to $\varepsilon_{Nd}$ values for F-poor biotite and two-mica granite from the western and eastern part of the Krušné Hory/Erzgebirge area [20]. In terms of their Nd isotope composition, the Cambro-Ordovician felsic and mafic volcanic rocks ($\varepsilon_{Nd}$ = −2.5 to +6) as well as the Proterozoic-Paleozoic metasediments of the Saxothuringian Zone ($\varepsilon_{Nd}$ = −6 to −11) [37, 38, 39] could be a potential source of both topaz granite groups. Past-published models involve the derivation of the Saxothuringian Hercynian granites by fractional crystallization of mantle derived melts [40] or by anatectic melting of entirely crustal sources [4, 6, 41, 42]. The above mentioned Nd isotope data supports the melting of crustal sources.

The Saxothuringian Late Precambrian to Lower Palaeozoic basement is distinctly heterogeneous and the Lower Palaeozoic mica-rich and feldspar-poor schist hosts high concentrations of HFSE, Sn and W [43]. On the other side, feldspar- and quartz-rich Late Precambrian metagreywackes offer a well suitable source of highly peraluminous and P-rich granite melt. The source composition of various granite melts could be also inferred from contents and/or ratios of some major elements (Al, Fe, Mg, Ti, Ca) in melts experimentally produced by dehydration melting of various lithologies [44, 45]. The most widely used for this purpose are the $CaO/Na_2O$ and $Al_2O_3/TiO_2$ ratios [45, 46] and/or the content in CaO, FeO, MgO and $TiO_2$ [44]. In both types of topaz granites, the low $CaO/Na_2O$ ratio is characteristic of pelite-derived peraluminous melts [46]. Using the $CaO/(FeO + MgO + TiO_2)$ ratio could discriminate two groups of variable ratios in the Saxothuringian topaz granites. High values of this ratio are characteristic of P-poor topaz granites, and consequently of granite melt produced by melting of felsic metapelites [44]. Other possible derivation of P-poor topaz granites could be small-volume partial melting of lower crustal felsic granulites [20]. This assumption is also supported by high content of REE and Y in these granites.

## CONTROLS OF PHOSHORUS ENRICHMENT

Contrasting patterns in the behavior of phosphorus in strongly peraluminous P-rich granites and weakly peraluminous P-poor granites could be explained by the strong control of peralumosity on apatite solubility in the granite melt [47, 48]. The significant enrichment in phosphorus of high-P granites leads to the formation of P- and Al-bearing complexes in residual melts resulting in high P-contents of feldspars, topaz and zircon. Phosphorus abundances together with phosphate mineralogy (absence of apatite but high amounts of monazite and/or xenotime in examined P-poor granites are opposite to high amounts of apatite and occurrence of phosphorus enriched zircon in examined P-rich granites) are widely used criteria to discriminate A-type from S-type topaz granites [2].

## ROLE OF FLUIDS

The origin and evolution of topaz granites is related to non-charge and non-ionic size (non-CHARAC) trace element behavior first described in detail by Bau [49]. There, the fractionation of high field strength elements (HFSE) is influenced by their complexation with fluids, especially with fluorine. According to Bau and Dulski [50] the complexation of HFSE with fluorine is marked by Y/Ho ratio > 28, while complexing with bicarbonate is assumed to generate values < 28. Thus, the value of Y/Ho is a highly valuable indicator of the fluid activity in fractionated topaz granite melts (F, $CO_2$). However, for P-poor topaz granites from the upper part of the Cínovec cupola the Y/Ho values are significantly low (18–20).

These granites also have a considerably high occurrence of REE fluorocarbonates. On the other hand, for all other examined P-poor topaz granites, as well as for P-rich granites, Y/Ho values are typically in the range of 29–40. The highly variable Y and Ho behavior in examined granites is reflected by the variable correlation between the Y/Ho ratio and the tetrad effect. Rare earth element fractionation of topaz granites, frequently described as the lanthanide tetrad effect [51, 52, 53], is another significant feature of topaz granites from the Krušné Hory/Erzgebirge area.

The REE tetrad effect was first discovered by Peppard et al. [54] in a system for liquid-liquid extraction of lanthanides. This effect causes a split of the chondrite normalized REE patterns into four consecutive curved segments that are referred to as tetrads, (first tetrad: La-Ce-Pr-Nd; second tetrad: Pm-Sm-Eu-Gd; third tetrad: Gd-Tb-Dy-Ho; fourth tetrad: Er-Tm-Yb-Lu). Equations by Irber [51] are usually used for quantification of the size of the tetrad effect. The highest values of the tetrad effect were found in P-poor granites in the upper part of the Cínovec cupola (up to 1.5), whereas P-rich granites from the Horní Slavkov–Krásno ore district show distinctly lower values of this effect (0.85–1.16). However, all analyzed P-rich topaz granites from the Krušné Hory/Erzgebirge area and P-poor granites from the upper part of the Cínovec stock show a positive correlation between Y/Ho and the value of the tetrad effect, whereas P-poor granites from the lower part of the Cínovec cupola show an opposite correlation of both parameters.

The most probable explanation of positive correlation between the REE-tetrad effect and Y/Ho ratio is high influence of F fluids on the fractionation of residual albite enriched melts. The influence of fluorine-enriched fluids is reflected by consecutive low-temperature albitization of these granites. This highly complicated treatment of late magmatic and postmagmatic highly aggressive fluids could be well documented by the epitaxic overgrowth of altered zircon by younger xenotime and the concurrent origin of REE fluorocarbonates in highly altered P–poor granites of the Cínovec cupola (Figure 9).

Figure 9. BSE image of epitaxial overgrowth of xenotime on zircon from topaz-bearing granite of the Cínovec stock (Zrn – zircon, Xtm – xenotime).

## CONCLUSION

The topaz granites evolved in the Central European part of the Hercynian granite belt, in particular in the Krušné Hory/Erzgebrige area, are highly fractionated members of a huge Hercynian magmatic activity in this area. The topaz-bearing granites form two distinct magmatic suites, which were emplaced in close proximity in this area. The first suite is represented by late-collisional, strongly peraluminous, P-rich, S-type granites. The second suite is formed by post-collisional slightly peraluminous, P-poor, A-type granites. Both suites display additional specific mineralogical and geochemical features. The P-rich suite contains a high amount of apatite, whereas the P-poor suite is absent of apatite. The P-rich suite has significant relatively low amounts of Zr, Y, Th and REE, while the P-poor suite has high contents of Y and Th, and chiefly HREE are typical. For discrimination of both granite suites, the most usable diagrams on S-type and A-type granites are based on content of Zr, Ce, Y and Th. Consequently, recommendation of the term A-granite being replaced by the term ferroan granite could be matter of additional discussion. Both granite groups originated with a somewhat similar highly mature protolith, distinctly enriched in lithophile elements (Li, Rb, Cs), Sn and W. However, P-rich granites display distinctly conspicuous features of metasedimentary protolith, whereas P-low granites show some features of meta-igneous protolith.

## ACKNOWLEDGMENTS

The whole-rock analysis was funded by the Czech Science Foundation (project No. 205/09/0540). Preparation of the paper also benefited from discussions with Prof. F. Holtz (Leibniz University, Hannover) in the frame of a joint research program supported by the Academy of Sciences of the Czech Republic and the German Science Foundation (project 436 TSE 113/48/0-1).

## REFERENCES

[1]  Taylor, R. P. *Can. Mineral.* 1992, 30, 895–921.
[2]  Bonin, B. *Lithos* 2007, 97: 1–29.
[3]  Breiter, K.; Förster, H-J.; Seltmann, R. *Mineral. Deposita* 1999, 34, 505–521.
[4]  Förster, H-J.; Tischendorf, G.; Trumbull, R. B.; Gottesmann, B. *J. Petrol.* 1999, 40, 1613–1645.
[5]  Förster, H-J.; Romer, R. L. In Pre-Mesozoic Geology of Saxo-Thuringia. Linnemann, U.; Romer, R. L.; Ed.; *Schweizerbart Science Publishers*: Stuttgart, 2010; pp 297–308.
[6]  Breiter, K. Lithos 2011, doi: 10.1016/j.lithos.2011.09.022.
[7]  Hösel, G.; Hoth, K.; Jung, D.; Leonhardt, D.; Mann, M.; Meyer, H.; Tägl, U. Das Zinnerz-Lagerstättengebiet Ehrenfriedersdorf/Erzgebirge; Bergbau in Sachsen, 1; Sächsische Landesamt für Umwelt und Geologie: Freiberg; 1994. pp 1–189.
[8]  René, M.; Škoda, R. *Mineral. Petrol.* 2011, 103, 37–48

[9] Breiter, K.; Frýda, J.; Seltmann, R.; Thomas, R. *J. Petrol*. 1997, 37, 74–94.
[10] Hösel, G.; Fritsch, E.; Josiger, U.; Wolf, P. (1996) Das Lagerstättengebiet Geyer; Bergbau in Sachsen, 4; Sächsisches Landesamt für Umwelt und Geologie: Freiberg, 1996; pp 1–112.
[11] Hoth, K.; Ossenkopf, W.; Hösel, G.; Zernke, B.; Eisenschimdt, K.; Kühne, R. *Geoprofil* 1991, 3, 3–13.
[12] Jarchovský, T. *J. Czech Geol. Soc*. 2006, 51, 201–216.
[13] René, M. *Acta Univ. Carol. Geol*. 1998, 42, 103–109,
[14] Müller, A.; René, M.; Behr, H-J.; Kronz, A. *Mineral. Petrol*. 2003, 79, 167–191.
[15] Förster, H-J.; Seltmann, R.; Tischendorf, G. High-fluorine, low phosphorus A-type (post-collision) silicic magmatism in the Erzgebirge. *Terra Nostra* 1995, 7, 32–35.
[16] Seifert, T.; Kempe, U. *Beih. Eur. J. Miner*. 1994, 6, 125–172.
[17] Dolejš, D.; Štemprok, M. *Bull. Czech Geol. Surv*. 2001, 76, 77–99.
[18] Štemprok, M.; Šulcek, Z. *Econ. Geol*. 1969, 64, 392–404.
[19] Förster, H-J.; Rhede, D. *N. Jb. Miner. Abh*. 2006, 182, 307–321.
[20] Breiter, K. *Z. geol. Wiss*. 2008, 36, 365–382.
[21] Breiter, K. *Bull. Czech Geol. Surv*. 2002, 77, 67–92.
[22] Chappel, B. W.; Hine, R. *Res. Geol*. 2006, 56, 203–244.
[23] Whalen, J. B.; Currie, K. L.; Chappell, B. W. *Contrib. Mineral. Petrol*. 1987, 95, 407–419.
[24] Chappel, B. W.; White, A. J. R. *Pac. Geol*. 1974, 8, 173–174.
[25] Clarke, D. B. Granitoid rocks; Topics in the Earth Sciences 7; Chapman and Hall: London, 1992; pp1–283.
[26] Loiselle, M. C.; Wones, D. R. Characteristic and origin of anorogenic granites. *Geol. Soc. Am. Abstr. Program* 1979, 11, 468.
[27] Collins, W. J.; Beams, D.; White, J. R.; Chappell, B. W. *Contrib. Mineral. Petrol*. 1982, 80, 189–200.
[28] Eby, G. N. *Lithos* 1990, 26, 115–134.
[29] Eby, G. N. *Geology* 1992, 20, 641–644.
[30] Goodman, R. J. *Geochim. Cosmochim. Acta* 1972, 36, 303–317.
[31] Cotton, F. A.;, Wilkinson, G. Advanced inorganic chemistry; *John Wiley Interscience*: New York, 1980.
[32] Creaser, R. A.; Price, R. C.; Wormald, R. J. *Geology* 1991, 19, 163–166.
[33] Frost, C. D.; Frost, B. R. *J. Petrol*. 2011, 52, 39–53.
[34] Taylor, R. P.; Fallick, A. E. *Terra Nova* 1997, 9, 105–108.
[35] Frost, B. R.; Arculus, R. J.; Barnes, C. G.; Collins, W. J.; Ellis, D. J.; Frost, C. D. *J. Petrol*. 2001, 42, 2033–2048.
[36] Romer, R. L.; Thomas, R.; Stein, H. J.; Rhede, D. *Miner. Deposita* 2007, 42, 337–359.
[37] Hegner, E.; Kröner, A. In Orogenic Processes: Quantification and Modelling in the Variscan Belt; Franke, W.; Haak, V.; Oecken, O.; Tanner, D.; Ed.; Geol. Soc. London, Spec. Publ. 179; *Geol. Soc. London*: London, 2000; pp 113–131.
[38] Linnemann, U.; Romer, R. L. *Tectonophysics* 2002, 352, 33–64.
[39] Linnemann, U.; McNaughton, N. J.; Romer, R. L.; Gehmlich, M.; Drost, K.; Tonk, C. *Int. J. Earth Sci*. 2004, 93, 683–705.

[40] Dahm, K. P. *Geol. Carpathica* 1985, 36: 397–410
[41] Štemprok, M. *Z. geol. Wiss.* 1993, 21, 237-245.
[42] Tischendorf, G.; Förster, H.-J. In *Mineral deposits of the Erzgebirge/Krušné Hory* (Germany/Czech Republic). Gehlen von, K.; Klemm, D.D.; Ed.; Monogr. Series on Mineral Deposits 31; Gebrüder Borntraeger: Berlin 1994, pp 5-23.
[43] Mingram, B. *Geol. Mag.* 1998, 135, 785-801.
[44] Patino Douce, A.E. In Understanding Granites. Castro, A.; Fernández, C.; Vigneresse, J. L.; Ed.; Special Publications 168; *Geological Society* London: London, 1999; pp. 55–75.
[45] Jung, S.; Pfänder, J. A. *Eur. J. Mineral.* 2007, 19, 859-870.
[46] Sylvester, P.J. *Lithos* 1998, 45, 29-44.
[47] Wolf, M.B.; London, D. *Geochim. Cosmochim. Acta* 1994, 58, 4127–4145.
[48] Chappel, B.W. *Lithos* 1999, 46, 535–551.
[49] Bau, M. Contrib. *Mineral. Petrol.* 1996, 123, 323-333.
[50] Bau, M.; Dulski, P. *Contrib. Mineral. Petrol.* 1995, 119, 213–223.
[51] Irber, W. *Geochim. Cosmochim. Acta* 1999, 63, 489-508.
[52] Monecke, T.; Dulski, P.; Kempe, U. *Geochim. Cosmochim. Acta* 2007, 71, 335–353.
[53] Monecke, T.; Kempe, U.; Monecke, J.; Sala, M.; Wolf D. *Geochim. Cosmochim. Acta* 2002, 66, 1185-1196.
[54] Peppard, D. F.; Mason, G. W.; Lewey, S. *Amer. Mineral.* 1969, 78, 1275–1285.
[55] Maniar, P. D.; Piccoli, P. M. *Geol. Soc. Amer. Bull.* 1989, 101, 635–643.
[56] Taylor, S. R.; McLennan, S. M. (1985). The continental crust: its composition and evolution. *Blackwell: Oxford,* 1985; pp 1–312.

In: Granite
Editors: Miroslava Blasik and Bogdashka Hanika
ISBN: 978-1-62081-566-3
© 2012 Nova Science Publishers, Inc.

*Chapter 6*

# NONLINEAR ACOUSTIC PHENOMENA IN GRANITE

## *V. E. Nazarov*[*], *A. B. Kolpakov, and A. V. Radostin*
Institute of Applied Physics, Russian Academy of Science, Russia

## ABSTRACT

Granite as a rock with rather complex microstructure possesses a strong acoustic nonlinearity in comparing with nonlinearity of homogeneous materials such as water, glass etc. This fact is associated with the presence in their structure of different defects such as dislocations, micro-cracks, grain boundaries etc, and it was confirmed and accepted by number of researchers. This chapter contains the results of two experimental studies of nonlinear acoustic phenomena in a bar resonator made of Pitkyarant Karelian granite. The first set of experiments is the study of amplitude dependent internal friction phenomena (nonlinear loss and shift in resonance frequency) and generation of the second harmonic in the case of bar exciting in the low-frequency range at its eigenfrequencies (from 3 kHz to 17 kHz). The revealed amplitude dependences are described analytically in the framework of hysteretic state equations analogously to Granato-Lücke dislocation theory. It was shown that cubic hysteresis manifests itself at small amplitudes of strains, whereas quadratic hysteresis occurs at large ones. In the second set of experiments the weak ultrasonic pulses (with carrier frequency in the range from 150 kHz to 1 MHz) were excited simultaneously with the low-frequency wave in resonance. It was revealed that characteristics of pulses propagation (the nonlinear attenuation and carrier frequency phase delay) are dependent on low frequency wave amplitude and frequency. The observed phenomena are described analytically in terms of the phenomenological state equation containing hysteretic and dissipative nonlinearities. Also a frequency dependence of nonlinearity is revealed. The effective values of nonlinear parameters of granite are estimated. The results of the study can be used for development of nonlinear acoustic diagnostics technique for rocks.

---

[*] Institute of Applied Physics, Russian Academy of Science, 46 Uljanov Str., Nizhny Novgorod, 603950, Russia, E-mail: *nazarov@hydro.appl.sci-nnov.ru.*

## INTRODUCTION

A research of nonlinear wave phenomena (NWP) in rocks is presently actual and of interest for many reasons. Anomalously high acoustic nonlinearity of such media (in comparison with homogeneous weak-nonlinear materials) is the most interesting of them[1-8]. From the "structural" point of view, rocks contain various kinds of nonlinear viscous-elastic defects (or micro-inhomogeneities): dislocations, borders of grains, cracks, etc., which differentiates them from homogeneous materials. Such media are called micro-inhomogeneous. The term micro-inhomogeneous[10,11] (or mesoscopic[4,5]) is used for medium containing defects that are greater than inter-atomic spacing but less than the wavelength. Also, it is assumed that there are many defects on the wavelength and their spatial distribution is statistically homogeneous. Thus, on average, it is possible to consider this medium as "acoustically homogeneous" or "macro-homogeneous" on the scale larger than the size of defects but less than the wavelength.

It is worth noting that traditional "classical" five-constant elasticity theory intended for weak-nonlinear homogeneous materials[7-9] can not explain the NWP regularities being observed in experiment with micro-inhomogeneous media and there is no "universal" theory adequately describing the NWP in such ones. Acoustic nonlinearity of micro-inhomogeneous solids is defined by defects in their structure. In this case, the dynamic state equations (i.e. the generalized dependences $\sigma = \sigma(\varepsilon)$, where $\sigma$ and $\varepsilon$ is stress and strain, respectively) of similar media even in the range of small strain amplitudes (which is typical for acoustic waves) often are not analytical (i.e. not smooth and not differentiable) and, as a rule, they have hysteresis in a low-frequency (LF) range.

As a rule, the acoustic nonlinearity of micro-inhomogeneous media contains a low-frequency (LF) hysteretic and high-frequency (HF) dissipative (inelastic) and/or reactive (elastic) components (their most important differences are connected with different dependences on amplitude and/or frequency of the acoustic action[2,3,12-16]). Specifically, the hysteretic nonlinearity decreases and dissipative nonlinearity increases at the frequency rise. The knowledge of this circumstance gives possibility to separate the contributions of LF hysteretic and HF dissipative (and reactive) nonlinearity to the manifestation of different phenomena and to carry out the experiments at the guaranteed dominating influence of this or that defect on the specific phenomenon studied. An experimental study of nonlinear phenomena allows to show up the mechanisms of the anomalous nonlinearity and to create the models of micro-inhomogeneous media. In turn, it is a basis for the development of diagnosis nonlinear acoustic methods that can control the structure and the state of these media. According to conventional notions, one of the causes of polycrystalline solids hysteretic nonlinearity (and the phenomena of amplitude-dependent internal friction (ADIF) connected with it) is moving of dislocations (one-dimensional defects of the crystal lattice)[17-21]. Experimental studies show that nonlinear acoustic phenomena manifest themselves differently in the different micro-inhomogeneous media, and this difference is more qualitative than quantitative. The reason of last circumstance is manifestation of especially individual nonlinear properties of such media and, as a consequence of it, an absence for them the uniform universal nonlinear state equation (which, generally speaking, is not deduced from the general thermodynamic principles as, for example, the equations of "classical" five-constant elasticity theory describing the deformation of weak-nonlinear (without defects) homogeneous materials[7-9]). From classical five-constant elasticity theory the reactive quadratic nonlinearity in the state equation follows and there is no dissipative nonlinearity. Nevertheless, for micro-inhomogeneous media, the nonlinear state equation can be reconstructed (at least, in general) on the basis of the analysis of the amplitude-frequency dependences of nonlinear phenomena revealed in experiments. In this case, experimental results possess the greatest "information value" for media with "nonlinear behavior" mismatching to the representations established traditionally.

The present chapter deals with the results of the experimental study of ADIF nonlinear phenomena, generation of LF wave second harmonic as well as interaction between LF and HF acoustic waves in a bar acoustic resonator made of granite. The analytical description of observed NWP is carried out within the framework of the phenomenological state equations containing the LF hysteretic and HF dissipative nonlinearities. From comparison of the calculation results with those of the experiment, the type of the hysteretic state equation and the values of effective nonlinearity parameters of granite were defined.

## EXPERIMENTAL SETUP

Experiments were carried out with the bar resonator made of Pitkyarant Karelian granite1: the length of the bar is L=39 cm, its section is a square with the side of 1.6 cm. Experimental measurements were carried out at room temperature. A block diagram of the experimental setup is shown on Figure 1. The resonator (1) was excited by a piezoceramic pump wave radiator (2) that cemented to a massive metallic loading (3) and to one bar end-wall. A piezoelectric radiator (4) for exciting of the ultrasonic pulses, and the piezoelectric accelerometer (5) for measurements of the pump wave amplitude, were cemented to the other (free) bar end-wall. In the vicinity of the pump radiator, an accelerometer (6), reacting to the longitudinal (i.e. along rod) component of acceleration, was cemented on the lateral bar surface and used to receive and measure the relative amplitudes of the ultrasonic pulses. The bar was considered as an acoustic resonator with almost rigid and soft boundaries.

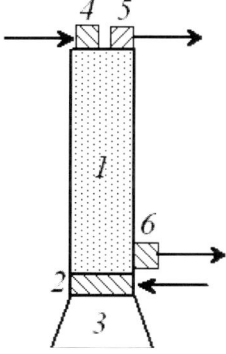

Figure 1. Schematic diagram of the experimental setup.

---

[1] Granite (Italian: granito, from Latin: granum — grain) is siliceous magmatic intrusive rock. Granite is mineral of volcanic nature and it consists from granules of the various size and color. It contains quartz, plagioclase, potash feldspar, micas – biotite and/or muscovite. The granite density is approximately 2.7 g/cm$^3$.

Its eigenfrequencies were proportional to odd numbers (1,3,5,…) and determined by the expression: $F_p \cong (2p-1)C_0/4L$, where $C_0$ is the phase velocity of longitudinal LF wave in the bar and $p$ is the mode number; $p = 1,2,3,\ldots$. Eigenfrequencies $F_p$ and $Q$-factors $Q_p$ of the resonator at low amplitude of its excitation, when nonlinear effects were not observed, have been determined by the traditional method, i.e. by using the resonance peaks of resonator and its $\Delta F_{0.71}$ range of resonance curve on $1/\sqrt{2} \cong 0.71$ level $Q_p \cong F_p/\Delta F_{0.71}$. The values $F_p$ and $Q_p$ for the first three longitudinal modes were: $F_1 \cong 3820$ Hz, $F_2 \cong 10220$ Hz, $F_3 \cong 17200$ Hz and $Q_1 \cong 255$, $Q_2 \cong 238$, $Q_3 \cong 273$. The eigenfrequency of the first mode corresponds to the phase velocity in the bar $C_0 \cong 5.3 \cdot 10^5$ cm/s.

## PHENOMENA OF *ADIF* AND SECOND HARMONIC HENERATION OF *LF* WAVE

In the first set of experiments LF harmonic oscillations were generated in the resonator at the frequency $F$ near the eigenfrequency $F_p$ of the first three longitudinal modes, and amplitude dependences of the nonlinear resonance frequency shift $\Delta F_{nl} = F - F_p < 0$, the nonlinear damping coefficient $\mu_{nl}$, and the amplitude of the second harmonic were measured. Figure 2 shows the dependences of strain amplitude $\varepsilon_m$ on the amplitude of voltage $U$ at the pumping radiator at the resonance oscillations of the bar. From this figure, it can be seen that the dependences $\varepsilon_m = \varepsilon_m(U)$ become nonlinear when the resonator excitation amplitude increases. It testifies to presence of amplitude-dependent losses. Figure3a, b shows the dependences of the relative module value of nonlinear resonance frequency shift $|\Delta F_{nl}/F_p|$ and relative nonlinear damping coefficient $\mu_{nl}/\mu_p$ on $\varepsilon_m$, where $\mu_p = (\Omega_p Q_p)^{-1}$, $\Omega_p = 2\pi F_p$, $p = 1,2,3$. The relative coefficient of nonlinear damping was defined by the formula $\frac{\mu_{nl}}{\mu_p} = \frac{\varepsilon_{m1}}{\varepsilon_m} \cdot \frac{U}{U_1} - 1$ describing a deviation of the observed dependence $\varepsilon_m = \varepsilon_m(U)$ (shown in Figure 1) from the linear dependence plotted through initial experimental points ($U_1$, $\varepsilon_{m1}$) corresponding to small amplitudes of resonator excitation when the ADIF effects are negligibly small and $\varepsilon_m \sim U$. As it is clearly seen from Figure 3, firstly, the $\varepsilon_m$ dependences of $|\Delta F_{nl}/F_p|$ and $\mu_{nl}/\mu_p$ do not depend on the frequency of excitation (for three first modes) and, secondly, two amplitude ranges can be distinguished in these dependences: first (I) range $\varepsilon_m < \varepsilon^* \cong 2 \cdot 10^{-6}$, where $\Delta F_{nl}/F_p \sim \varepsilon_m^2$,

$\mu_{nl}/\mu_p \sim \varepsilon_m^2$, and second (II) range $\varepsilon_m > \varepsilon^*$, where (with the exception of one-two last points for nonlinear loss) $\Delta F_{nl}/F_p \sim \varepsilon_m$, $\mu_{nl}/\mu_p \sim \varepsilon_m$. Obviously that, for each range, the relation of elasticity nonlinear modulus defect to damping decrement, i.e. parameter

$$r_{1,2} = \left(\frac{|\Delta F_{nl}/F_p|}{2\pi F_p \mu_{nl}}\right)_{1,2},$$ does not depend from amplitude $\varepsilon_m$. In this case, $r_1 \cong 4.7$ and $r_2 \cong 3.3$. The presence of identical amplitude dependences of the shift in the resonant frequency and nonlinear loss in each range testifies to the manifestation of hysteretic power nonlinearity (with exponent $n$) in the resonator material (for the first range $n=3$, and for the second range $n=2$). In addition at sufficiently strong excitation of the resonator (at $\varepsilon_m \geq 3\cdot 10^{-6} > \varepsilon^*$, i.e. in the second range), generation of a second harmonic of the pumping wave was also observed. Figure 4 shows the dependence of displacement amplitude $W_2(L)$ of the free bar end-wall at the second harmonic frequency on amplitude of strain $\varepsilon_m$ (in exact resonance) when pumping wave was excited at the first mode. It can be seen from this figure that the $\varepsilon_m$ dependence $W_2(L)$ is close to the quadratic $W_2(L) \sim \varepsilon_m^2$, that also testifies (in the second range) to the manifestation of quadratic hysteretic nonlinearity of the considered granite sample.

An analytical description of ADIF phenomena and generation of the second harmonic we will carry out within the framework of the phenomenological state equation containing an elastic hysteresis[3], which is similar to dislocation one proposed by Granato and Lücke[17]:

$$\sigma(\varepsilon, \operatorname{sgn}\dot\varepsilon, \dot\varepsilon) = E[\varepsilon - f(\varepsilon, \operatorname{sgn}\dot\varepsilon)] + \alpha\rho\dot\varepsilon, \qquad (1)$$

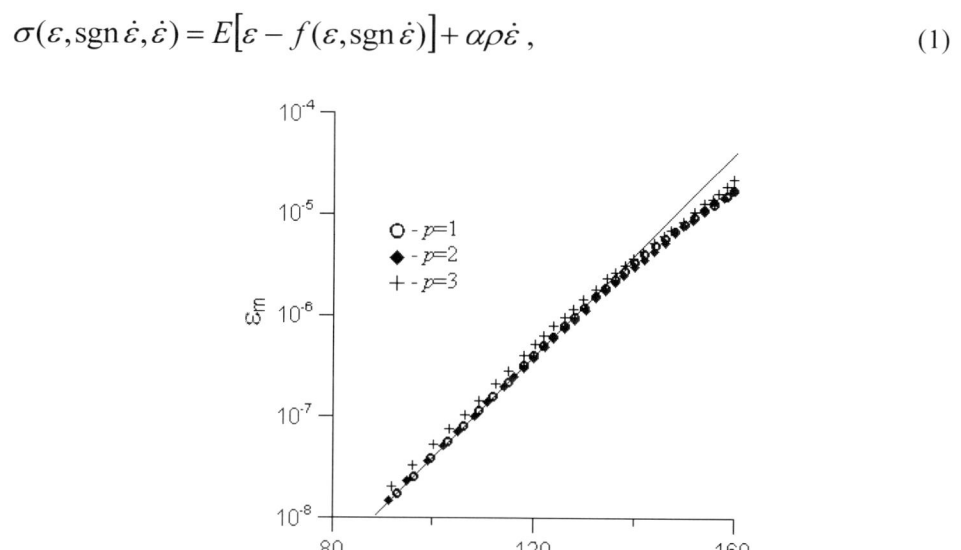

Figure 2. Wave amplitude $\varepsilon_m$ in resonator versus voltage $U$ at the pumping radiator (in dB relative to 1 $\mu V$). The straight line corresponds to the linear dependence: $\varepsilon_m \sim U$.

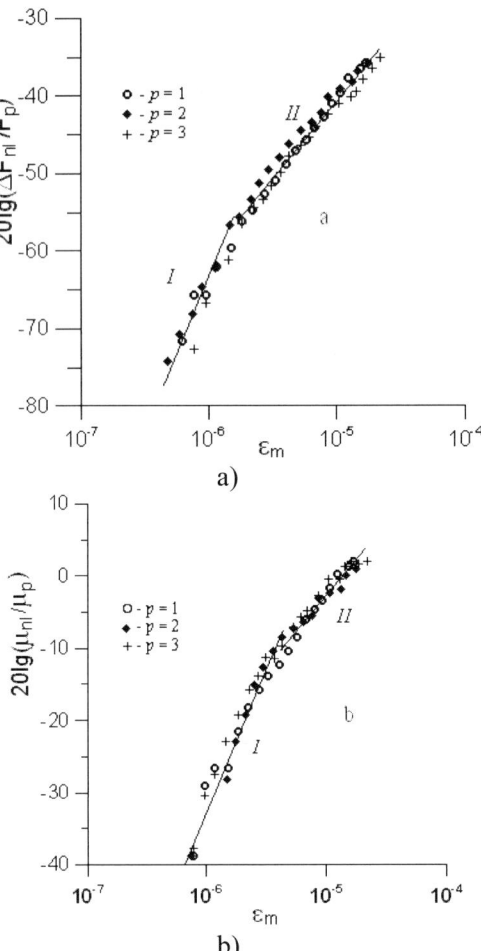

Figure 3. Relative nonlinear shift of resonance frequency (a) and damping coefficient (b) versus wave amplitude in resonance. The straight lines correspond to dependences: I - $\Delta F_{nl} / F_1 \sim \varepsilon_m^2$, $\mu_{nl} / \mu_1 \sim \varepsilon_m^2$, II - $\Delta F_{nl} / F_1 \sim \varepsilon_m$, $\mu_{nl} / \mu_1 \sim \varepsilon_m$.

$$f(\varepsilon, \operatorname{sgn} \dot{\varepsilon}) = \frac{1}{n} \begin{cases} \gamma_1 \varepsilon^n, & \varepsilon > 0, \dot{\varepsilon} > 0; \\ (\gamma_1 + \gamma_2) \in_m^{n-1} \varepsilon - \gamma_2 \varepsilon^n, & \varepsilon > 0, \dot{\varepsilon} < 0; \\ -\gamma_3 \varepsilon^n, & \varepsilon < 0, \dot{\varepsilon} < 0; \\ (-1)^n (\gamma_3 + \gamma_4) \in_m^{n-1} \varepsilon + \gamma_4 \varepsilon^n, & \varepsilon < 0, \dot{\varepsilon} > 0, \end{cases} \quad (2)$$

where $\sigma$, $\varepsilon$ and $\dot{\varepsilon}$ - are longitudinal stress, strain, and strain rate, respectively; $E$ is Young's modulus; $f = f(\varepsilon, \operatorname{sgn} \dot{\varepsilon})$ is hysteretic function, $\left| \frac{\partial f(\varepsilon, \operatorname{sgn} \dot{\varepsilon})}{\partial \varepsilon} \right| \ll 1$, $\gamma_{1-4}$ are parameters of hysteretic nonlinearity; $\in_m = \in_m (x)$ is local strain amplitude, $\in_m < |\in_{th}|$, $\in_{th}$ is yield strain (when it is exceeded, irreversible plastic strains appear in the solid and it is broken; for many materials $|\epsilon_{th}| > 10^{-4} - 10^{-3}$); $\alpha$ is viscosity coefficient ($\alpha = C_0^2/\Omega_p Q_p$), $\rho$ is density; $n = 3$ for the first range and $n = 2$ for the second range. $|\gamma_{1-4} \epsilon_m^{n-1}| \ll 1$, $|\gamma_{1-4}| \gg 1$. (Here and further, we use the same indexation of parameters $\gamma_i$ and at $\varepsilon_m < \varepsilon^*$ and $\varepsilon_m > \varepsilon^*$, but, of course, the values of these parameters are different in different ranges).

Figure5 displays the qualitative view of quasi-static hysteresis (1), (2) (when $\alpha \rho |\dot{\varepsilon}| \ll E|f(\varepsilon, \text{sgn}\, \dot{\varepsilon})|$).

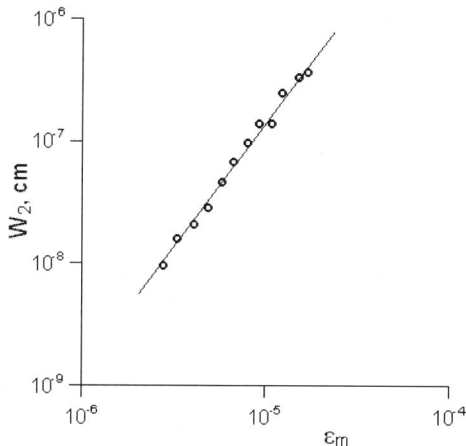

Figure 4. The amplitude of displacement $W_2(L)$ of the free end-wall of the bar at the frequency of the second harmonic versus amplitude $\varepsilon_m$ (in resonance) at resonator excitation at the frequency of the first mode. The straight line corresponds to dependence: $W_2 \sim \varepsilon_m^2$.

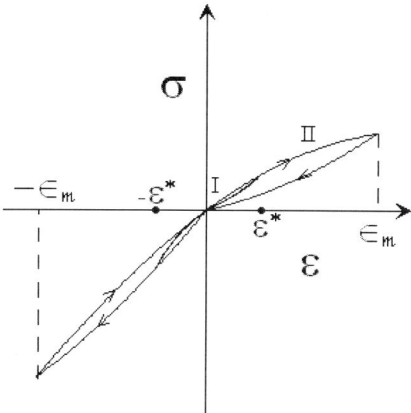

Figure 5. Quasi-static dependences $\sigma = \sigma(\varepsilon, \text{sgn}\,\dot{\varepsilon}, \dot{\varepsilon})$ for the hysteresis (1) and (2) at small ($\varepsilon_m < \varepsilon^*$ - I) and great ($\varepsilon_m > \varepsilon^*$ – II) strain amplitudes.

The state Eq. (1), (2), together with the equation of motion $\rho W_{tt} = \sigma_x(\varepsilon, \operatorname{sgn}\dot\varepsilon, \dot\varepsilon)$ and boundary conditions at the end-walls of the resonator: $W(x=0,t) = A\cos\Omega t$, $W_x(x=L,t)=0$ (where $W = W(x,t)$ is displacement, $\varepsilon(x,t) = \dfrac{\partial W(x,t)}{\partial x}$; $A$ and $\Omega$ is amplitude ($A \sim U$) and frequency of the pumping radiator, respectively) describe LF nonlinear wave processes in such a resonator. Their calculation was performed on the basis of the perturbation method because the following conditions were satisfied in the experiment $|\Delta F_{nl}/F_p| \ll 1$, $\varepsilon_2 \ll \varepsilon_1 \cong \varepsilon_m$, ($\varepsilon_2$ is strain amplitude at frequency $2\Omega$). The frequency response of the resonator is determined by expression

$$\varepsilon_m = \frac{A_0 \Omega_p}{L} \frac{1}{[(\delta + \delta_{nl})^2 + (\mu_p + \mu_{nl})^2 \Omega_p^4 / 4]^{1/2}}, \qquad (3)$$

where the nonlinear shift in resonance frequency $\delta_{nl} = 2\pi \Delta F_{nl}$ and the nonlinear damping coefficient $\mu_{nl}$ in the first and second ranges depend on parameters $\gamma_i$ of the hysteresis (2), $\delta = \Omega - \Omega_p$, $|\delta| \ll \Omega_p / p$. In the first range ($\varepsilon_m < \varepsilon^*$), expressions for $\Delta F_{nl}/F_p$ and $\mu_{nl}/\mu_p$ are written as

$$\Delta F_{nl}/F_p = -a_1 \varepsilon_m^2, \quad \mu_{nl}/\mu_p = b_1 Q_p \varepsilon_m^2, \qquad (4)$$

where $a_1 = \dfrac{1}{32}\left\{(\gamma_1 + \gamma_2 - \gamma_3 - \gamma_4) + \dfrac{3}{4}(\gamma_1 - \gamma_2 - \gamma_3 + \gamma_4)\right\}$, $b_1 = \dfrac{\gamma_1 + \gamma_2 - \gamma_3 - \gamma_4}{16\pi} > 0$.

The comparison of the experimental results (Figure 3a, b) and Eq. (4) allows us to find the values of coefficients $a_1$ and $b_1$, and parameters $\gamma_1 - \gamma_3$, $\gamma_2 - \gamma_4$: $a_1 = 5.6 \cdot 10^8$, $b_1 = 1.2 \cdot 10^8$, $\gamma_1 - \gamma_3 = 1.1 \cdot 10^{10}$, $\gamma_2 - \gamma_4 = -4.9 \cdot 10^9$. From the first expression (4), it is possible to estimate a value of the effective parameter $\Gamma_3$ of granite cubic nonlinearity (at small amplitudes $\varepsilon_m < \varepsilon^*$). Accepting $n = 3$, $\gamma_1 + \gamma_2 = 0$, $\gamma_3 + \gamma_4 = 0$ and $\gamma_1 = -\gamma_3 = \Gamma_3$ in equation (2), we find the state equation (1) containing the elastic cubic nonlinearity $f(\varepsilon) = \Gamma_3 \varepsilon^3 / 3$ only, where $\Gamma_3 = 32 a_1 / 3 = 6 \cdot 10^9$.

In the second range ($\varepsilon_m > \varepsilon^*$), expressions for $\Delta F_{nl}/F_p$ and $\mu_{nl}/\mu_p$, as well as for amplitude $W_2(L)$ oscillations at the frequency of the second harmonic have the form

$$\Delta F_{nl}/F_p = -a_2 \varepsilon_m, \quad \mu_{nl}/\mu_p = b_2 Q_p \varepsilon_m, \qquad (5)$$

$$W_2(L) = \sqrt{a_2^2 + b_2^2}\, \varepsilon_m^2 L, \qquad (6)$$

where $a_1 = \dfrac{4}{9\pi^2}(\gamma_1 - \gamma_2 + \gamma_3 - \gamma_4) + \dfrac{1}{6\pi}(\gamma_1 + \gamma_2 + \gamma_3 + \gamma_4)$,

$b_1 = \dfrac{2}{9\pi^2}(\gamma_1 + \gamma_2 + \gamma_3 + \gamma_4) > 0$,

$a_2 = \dfrac{1}{64}(\gamma_1 - \gamma_2 - \gamma_3 + \gamma_4) + \dfrac{1}{24\pi}(\gamma_1 + \gamma_2 - \gamma_3 - \gamma_4), b_2 = \dfrac{1}{48\pi}(\gamma_1 + \gamma_2 - \gamma_3 - \gamma_4)$.

By comparing of the experimental results (Figs.3, 4) and the expressions (5), (6), (7) we find coefficients $a_1$, $b_1$, $\sqrt{a_2^2 + b_2^2}$ and corresponding parameters $\gamma_1 + \gamma_3$ and $\gamma_2 + \gamma_4$: $a_1 = 9 \cdot 10^2$, $b_1 = 2.7 \cdot 10^2$, $\gamma_1 + \gamma_3 = 8.9 \cdot 10^3$, $\gamma_2 + \gamma_4 = 3.3 \cdot 10^3$, $\sqrt{a_2^2 + b_2^2} = 30$. From expression (6) it is possible to estimate the effective parameter $\Gamma_2$ of granite quadratic elastic nonlinearity (at $\varepsilon_m > \varepsilon^*$). By accepting, in the equation (2) $n = 2$, $\gamma_1 + \gamma_2 = 0$, $\gamma_3 + \gamma_4 = 0$ and $\gamma_1 = -\gamma_3 = \Gamma_2$, we obtain the state equation (1) containing the elastic quadratic nonlinearity $f(\varepsilon) = \Gamma_2 \varepsilon^2 / 2$ only, where $\Gamma_2 = 16\sqrt{a_2^2 + b_2^2} = 480$. As one would expect, for the granite sample considered in this chapter, the $\Gamma_2$ value is $10^2$ times greater than the similar parameter of relatively "rigid" solids (a steel, glass)[7,8].

## A DAMPING AND A PHASE DELAY OF A CARRIER FREQUENCY OF WEAK ULTRASONIC PULSES UNDER THE ACTION OF POWERFUL LF WAVE

In the second set of experiments weak ultrasonic pulses were emitted by the HF radiator (4) simultaneously with excitation of resonant LF pump wave. After passage through the bar, pulses were accepted by accelerometer (6) and supplied on a digital oscilloscope-spectrum analyzer where measurements of their amplitude $U(\varepsilon_m)$ and the carrying frequency phase delay $\Delta\tau(\varepsilon_m)$ were carried out. The characteristics of the pulses passed through the bar were found to be dependent on strain amplitude $\varepsilon_m$ of powerful resonant LF wave. The pulses had duration $\tau = 60$ $\mu s$, their carrying frequency $f$ was in a range from 150 kHz to 1 MHz and repetition frequency was equal 26 Hz. The velocity $C$ of ultrasonic pulses propagation in the bar defined by their delay (when $\varepsilon_m = 0$) was $5 \cdot 10^5$ cm/s approximately.

An estimation of pulses length gives $l = C\tau = 30$ cm; this allows assuming pulses propagation in the bar as it was in a boundless medium.

When the strain amplitude $\varepsilon_m$ of the LF pump wave (at $\varepsilon_m > 10^{-6} > \varepsilon^*$, i.e., in the second range) increases, the noticeable decrease of the received amplitudes $U(\varepsilon_m)$ of pulses and increase of their phase delay $\Delta\tau(\varepsilon_m)$ were observed.

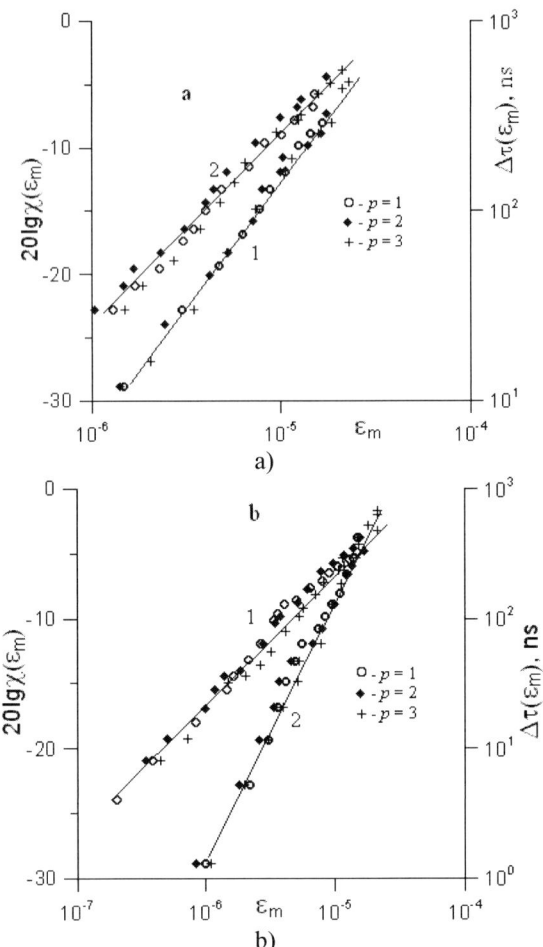

Figure 6. Nonlinear damping coefficient (1) and phase delay (2) for pulses with frequencies $f = 277$ kHz (a) и $f = 490$ kHz (b) versus LF wave strain amplitude $\varepsilon_m$ (in resonance) at resonator excitation at the frequencies of the three modes. The straight lines correspond to dependences: $\chi(\varepsilon_m) \sim \varepsilon_m$, $\Delta\tau(\varepsilon_m) \sim \varepsilon_m$.

Figure 6a, b present the $\varepsilon_m$ dependencies of nonlinear damping coefficient $\chi(\varepsilon_m) = \ln[U_0 / U(\varepsilon_m)]$ (where $U_0$ is the amplitude of the pulse without the pump wave) and the phase delay of pulses of $f = 277$ kHz and $f = 490$ kHz at the bar resonance excitation at its first three eigenfrequencies. It is seen from these plots that $\chi(\varepsilon_m)$ and $\Delta\tau(\varepsilon_m)$ are proportional to the strain amplitude $\varepsilon_m$ of LF pump wave and do not depend from its frequency $F_p$, i.e. $\chi(\varepsilon_m) \sim \varepsilon_m$, $\Delta\tau(\varepsilon_m) \sim \varepsilon_m$. Figure7 illustrates the dependences of $\chi = \chi(\varepsilon_m)$ and $\Delta\tau(\varepsilon_m)$ as the functions of pulse frequency $f$ at the excitation of resonator on its second mode ($p = 2$) frequency when $\varepsilon_m = 10^{-5}$. As it follows from this figure, basically, the coefficient $\chi = \chi(\varepsilon_m)$ increases [$\chi(\varepsilon_m) \sim f$] with a rise of $f$. And the phase delay ($\Delta\tau(\varepsilon_m)$) decreases considerably [$\Delta\tau(\varepsilon_m) \sim f^{-1}$] from 150 kHz to 400 kHz and $\Delta\tau(\varepsilon_m) \approx$ const in the higher frequency range (from 400 kHz to 1 MHz). Such $f$ dependencies $\chi(\varepsilon_m)$ and $\Delta\tau(\varepsilon_m)$ testify about the frequency relation (i.e. dispersion) of the nonlinear acoustic properties of granite[12-16].

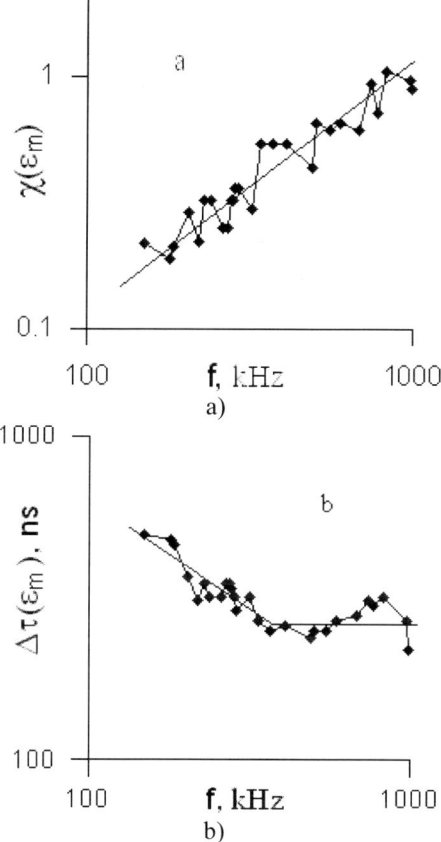

Figure 7. Nonlinear damping coefficient (a) and phase delay (b) of carrier frequency versus the pulses frequency $f$ at resonator excitation at the frequency of the second mode ($p = 2$), at $\varepsilon_m = 10^{-5}$. The straight lines correspond to dependences: $\chi(\varepsilon_m) \sim f_m$, $\Delta\tau(\varepsilon_m) \sim f^{-1}$, $\Delta\tau(\varepsilon_m) \approx$ const.

Analytical description of nonlinear damping phenomena and phase delay of the carrier frequency of weak ultrasonic pulses we will carry out within the framework of the phenomenological state equation containing the LF hysteretic (2) and HF dissipative nonlinearity:

$$\sigma(\varepsilon,\dot\varepsilon) = E[\varepsilon - f(\varepsilon,\dot\varepsilon)] + \alpha\rho[1 + \delta|\varepsilon|]\dot\varepsilon, \qquad (7)$$

where $\delta$ is the dimensionless parameter of dissipative nonlinearity. (Here, it is necessary to take into account that the parameters $\gamma_i$, $\alpha$ and $\delta$ depend on the acoustic pulse frequency

$\omega$ so as the granite possesses a nonlinearity dispersion in the HF range). In this case, the pulse nonlinear damping coefficient $\chi(\varepsilon_m)$, being caused by dissipative nonlinearity $\alpha\rho\delta|\varepsilon|\dot{\varepsilon}$, can be determined by the expression:

$$\chi(\varepsilon_m) = \frac{2\alpha\delta\varepsilon_m\omega^2 L}{\pi^2 C^3}, \qquad (8)$$

and the phase delay $\Delta\tau_h(\varepsilon_m)$ being caused by the hysteretic nonlinearity (2):

$$\Delta\tau_h(\varepsilon_m) = \frac{\gamma_0}{2\pi^2}\frac{\varepsilon_m L}{C}, \qquad (9)$$

where $\gamma_0 = \gamma_1 - \gamma_2 + \gamma_3 - \gamma_4 + \frac{\pi(\gamma_1 + \gamma_2 + \gamma_3 + \gamma_4)}{4}$.

From comparison between these expressions and experimental results, we determine effective parameters of dissipative and hysteretic nonlinearity of granite for $f_0 = 150$ kHz (when $\alpha\omega_0^2 L/2C^3 \cong 1$): $\delta \cong 6.2\cdot 10^4$ and $\gamma_0 \cong 1.4\cdot 10^4$. (The $\gamma_0$ parameter value determined by using the phase delay of the ultrasonic pulse carrier frequency $f_0 = 150$ kHz has appeared close to value determined by the HF phenomena of ADIF: $\gamma_0 \cong 1.5\cdot 10^4$) From Figure 7, it is seen that the coefficient $\chi(\varepsilon_m) \sim \alpha(f)\delta(f)f^2$ increases (in average) as $f/f_0$, i.e. $\alpha(f)\delta(f) \sim f^{-1}$ at the frequency increase (from 150 kHz to 1 MHz) and the hysteretic nonlinearity parameter $\gamma_0(f)$ decreases as $f_0/f$ in range from 150 kHz to 400 kHz and $\gamma_0 \cong 8\cdot 10^3$ at the more high frequencies (from 400 kHz to 1 MHz).

## CONCLUSION

The experimental and theoretical studies of the nonlinear acoustic phenomena in the resonator made of Karelian granite have shown that in the LF range (at least to 17 kHz); the granite acoustic nonlinearity is hysteretic and independent on frequency. In this case, the cubic and quadratic (on deformation) hysteresis are observed at the small ($\varepsilon_m < \varepsilon^* \cong 2\cdot 10^{-6}$) and great ($\varepsilon^* < \varepsilon_m < 3\cdot 10^{-5}$) strain amplitudes, respectively. The effective parameters of cubic (when $\varepsilon_m < \varepsilon^*$) and quadratic (when $\varepsilon^* < \varepsilon_m < 3\cdot 10^{-5}$) elastic nonlinearity exceed several orders of magnitude (in the case of the quadratic nonlinearity parameter $10^2$ times) of the analogous parameters of the relatively "rigid" homogeneous materials. For example, the magnitude of the parameter $\Gamma_3$ for the glass[7] is $6 \cdot 10^9$ and $\Gamma_2$ for the steel[8] is $480$.

In the range from 150 kHz to 1 MHz, Pitkyarant Karelian granite acoustic nonlinearity is anomalous high, frequency-independent, and containing both dissipative and hysteretic components. In this case, in general, the sound-by-sound damping coefficient being caused by dissipative nonlinearity increases and the carrier frequency HF pulse phase delay that connected with hysteretic nonlinearity decreases.

Hysteretic and dissipative nonlinearity of the granite give the possibility to use the nonlinear acoustic phenomena for the diagnostic control of rocks in the earth stratum.

## ACKNOWLEDGMENTS

This work was supported by Russian Foundation for Basic Research.

## REFERENCES

[1] Nazarov, V.E., Ostrovsky, L.A., Soustova, I.A., Sutin, A.M. *Physics of the Earth and Planetary Interiors*. 1988, *50*, 65-73.
[2] Nazarov, V.E. *Phys. Metals and Metallography*, 1991, *71*, 171-177.
[3] Zimenkov, S.V., Nazarov, V.E., *Izv. Acad. Sci. USSR Phys. Solid Earth*, 1993, 12-18. [in Russian].
[4] Guyer, R.A., Johnson, P.A. *Physics Today*. 1999, 30-36.
[5] Ostrovsky, L.A., Johnson, P.A. *La Rivista del Nuovo Cimento*, 2001, *24*, 1-46.
[6] Rudenko, O.V. *Physics-Uspekhi* 2006, *49*, 69-87.
[7] Zarembo, L.K., Krasil'nikov, V.A. *Sov. Phys. Usp.* 1971, *13*, 778–797.
[8] Zarembo, L.K., Krasil'nikov, V.A. *Introduction to Nonlinear Acoustics*; Nauka: Moscow, 1966. [in Russian].
[9] Landau, L.D., Lifshitz, E.M. *Theory of Elasticity*; Pergamon Press: New York, 1986.
[10] Isakovich, M.A. *Sov. Phys. Usp.*, 1979, *22*, 928–933.
[11] Isakovich, M.A., *General acoustics*; Nauka: Moscow, 1973. [in Russian].
[12] Nazarov, V.E., Radostin, A.V. *Acoust. Phys.* 2004, *50*, 446-453.
[13] Nazarov, V.E., Radostin, A.V. *Acoust. Phys.* 2005, *51*, 280-285.
[14] Nazarov, V.E., Kolpakov, A.B., Radostin, A.V. *Acoust. Phys.* 2007, *53*, 254-263.
[15] Nazarov, V.E., Kolpakov, A.B., Radostin, A.V. *Acoust. Phys.* 2009, *55*, 100-107.
[16] Nazarov, V.E., Kolpakov, A.B., Radostin, A.V. *Acoust. Phys.* 2010, *56*, 453-456.
[17] Granato, A., Lücke, K. *J. Appl. Phys.* 1956, *27*, 583-593.
[18] Niblett, D., Wilks, J. *Advances in Physics*. 1960, 9, 1–88.

[19] Mason, W. ed. *Physical Acoustics and Methods*, Vol. 4, Part A; Academic Press: New York and London, 1966.
[20] Truell, R., Elbaum, C., Chick, B.B. *Ultrasonic Methods in Solid State Physics*, Academic Press: New York and London. 1969.

[21] Asano, S. *J. Phys. Soc. Jap.* 1970, 29, 952-963.

In: Granite
Editors: Miroslava Blasik and Bogdashka Hanika
ISBN: 978-1-62081-566-3
© 2012 Nova Science Publishers, Inc.

*Chapter 7*

# EXPERIMENTAL METHODS OF DETERMINING THERMAL PROPERTIES OF GRANITE

*Olukayode D. Akinyemi,*[1,*]
*Thomas J. Sauer*[2]*, and Yemi S. Onifade*[3]

[1]Department of Physics, Federal University of Agriculture, Abeokuta, Nigeria
[2]National Laboratory for Agriculture and the Environment, Ames, Iowa US
[3]Department of Physics, Federal University of Petroleum Resources, Effurun, Nigeria

## ABSTRACT

Determination of thermal properties of granite using the block method is discussed and compared with other methods. Problems that limit the accuracy of contact method in determining thermal properties of porous media are evaluated. Thermal properties of granite are determined in the laboratory with and without application of thermal interface material TIM (Arctic Silver®) to study effects of thermal contact resistance. The thermal properties analyzer KD2 (line - source heat dissipation probe) is also used with and without TIM to measure thermal conductivity of the sample. Results from block method and KD2 analyzer, with and without TIM, are compared with standard values. Results indicate significant differences with consideration of thermal contact resistance. Thermal conductivity of the granite sample increased from 2.95 W/mK to 3.95 W/mK with the standard values (Kappelmayer and Haenel, 1974) for granite ranging from 2.0 and 7.0 W/mK. Volumetric heat capacity decreased from $7.17 \times 10^4$ J/m$^3$K to $5.95 \times 10^4$ J/m$^3$K , thermal diffusivity increased from $0.412 \times 10^{-4}$ m$^2$/s to $0.67 \times 10^{-4}$ m$^2$/s while heat flux density increased from $2.63 \times 10^{-1}$ W/m$^2$ to $4.9 \times 10^{-1}$ W/m$^2$. The difference in thermal conductivities with or without TIM is significant at ($P > 0.05$) which implies the effectiveness of the thermal interface material in reducing thermal contact resistance.

**Keywords:** Porous Media, Thermal properties, Block Method, Thermal Contact Resistance, Granite

---

[*] akinyemi@physics.unaab.edu.ng.

# 1. INTRODUCTION

Thermal properties (Thermal conductivity, Volumetric heat capacity, Thermal diffusivity, and heat flux density) of granite are important material properties in the study of underground heat transfer, and thus determining the value is vital in correctly modeling rock thermal behavior. Thermal properties of porous media are of great importance to environmental sciences, agriculture and engineering, especially in relation to temperature and heat flux in the root and rock zones [1]. Thermal and physical properties were measured on 206 Jurassic granite samples obtained from three boreholes in the central part of Korea and experimental data show that thermal conductivity increases with increasing thermal diffusivity and heat capacity [2]. Thermal properties of rock materials are also important in understanding energy, water balance and mass exchange processes occurring across porous media surfaces. Contact resistance error has been the greatest concern with regards to accuracy of thermal properties measurements [3–7].

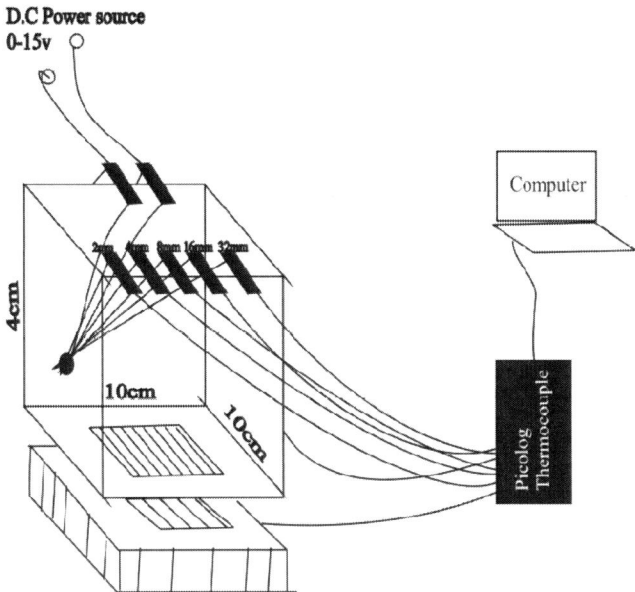

Figure 1. Contact Block Setup.

The connection between thermal properties and moisture content of rocks obtained during laboratory experiments have also been used to determine these properties in the field. Thermal parameters depend mainly on rock constituents, porosity and moisture content. Heat capacity is a material property, which expresses the fact that for changing the temperature of a certain volume of material, energy flows in or out and usually linked to the density of the material. When dynamic processes are involved, change of temperature versus time, at known boundary conditions is defined by both thermal conductivity and heat capacity. Block method, which employs a fabricated block of an appropriate material at a uniform temperature, had been used in the past to determine thermal properties of porous materials. The technique had largely been abandoned due to accuracy concerns arising from thermal contact resistance between the block and the sample surface. Investigators have also made attempts to measure thermal properties of rock materials using other methods [8-15], but very little is known on the use of block method to determine thermal properties of rocks. The objective of this work was therefore to discuss measurement of thermal properties of granite with special emphasis on block method. Efforts were also made to address the limitations of block method using heat sink compounds.

## 2. BLOCK METHOD

The method makes use of a solution of the heat conduction equation for two bodies of different temperature conditions suddenly brought into contact. From the temperature of the contact plane recorded during the measurement, the thermal properties of the homogenous body can be calculated if the thermal properties of the other body (a block of appropriate material placed on the sample) is known).

### 2.1. Method and Instrumentation

Block method device was fabricated from Perspex (10 x 10 x 4 cm) with $\lambda_p$ = 0.18568 W/mK, $C_p$ = 1.728 x $10^{-4}$ J/m³K. Copper - constantan thermocouples line the flat surface and at several depths (2, 4, 8, 16 and 32 mm) inside the block (Fig 1), at which the initial temperature at the instant t = 0, which must be uniform, is measured. The device obtains measurements of thermal properties at the surface which no other method that use line source probes does including KD2 thermal properties analyzer. The hypothesis is that since the material of interest was uniform with depth, the block and the KD2 data should agree, however many times in soil, the surface is either much wetter or drier than below so that the block method is really the only way to get the true surface layer thermal properties. The block with an insulation cover (2.54 cm - thick Styrofoam) is placed in a thermostat and after a few hours, the temperature at the surface of the block and within it was recorded for a short time to measure the initial temperature of the block and ensure a uniform temperature. After removing the insulation plate covering the lower surface, the block is immediately placed on the sample surface, the time of contact being taken at t = 0, while the contact temperature is logged for about 5 minutes.

The temperature near the center of the contact plane is calculated from the theory of two bodies suddenly brought into contact along the plane z = 0 at the instant t = 0, the temperature changes according to the equation [16]:

$$\frac{\delta \theta_i(z,t)}{\delta t} = a_i \frac{\delta^2 \theta_i(z,t)}{\delta z^2} \tag{1}$$

with i = 1 for block, and i = 2 for sample, thermal diffusivity a (m²/sec) = $\lambda$/ C = $\lambda$/pc, where $\lambda$ (W/mK) is the thermal conductivity, C (J/m³ °C) is the heat capacity per unit mass, p (kg/m³) is the density, and C (J/ kg °C) is the specific heat. Using the Laplace Transform of $\theta_1$(z, t), the general solution is given as [6]:

$$L\{\theta_i(z,t)\} = A_i \exp\left(-z\sqrt{\frac{p}{a_i}}\right) + B_i \exp\left(+z\sqrt{\frac{p}{a_i}}\right) + S_i(z,p) \qquad (2)$$

Calculations made by Van Wijk [6] and elaborated by Stigter [7] based on solving the Fourier equation for two finite bodies having different initial temperatures and brought at time t = 0 in contact at plane z = 0 have been repeated in this work. The solution for the temperature of the block's contact plane is given as:

$$\theta_1(0,t) = \frac{T_{1in}\sqrt{\lambda_1 C_1} + T_{2in}\sqrt{\lambda_2 C_2}}{\sqrt{\lambda_1 C_1} + \sqrt{\lambda_2 C_2}} + \frac{2}{\pi}\frac{\lambda_1 E_1 + \lambda_2 E_2}{\sqrt{\lambda_1 C_1} + \sqrt{\lambda_2 C_2}}\sqrt{t} \qquad (3)$$

where $T_{1in}$ is initial surface temperature of the block.

Using equation 3, and introducing

$$\alpha = \frac{\sqrt{\lambda_2 C_2}}{\sqrt{\lambda_1 C_1}},$$

then $\theta_{1(0,t)} = \frac{T_{1in} + \alpha T_{2in}}{1+\alpha}$

From equation 3, a plot of $T_1(0,t)$ vs. $\sqrt{t}$ yields a straight line graph with intersect $T_1(0,0)$ at t = 0. Using temperature readings from Block (1) and that from granite (2), a set of two equations is generated which can be solved to determine the surface temperature of the porous medium at t = 0. The porous medium temperature $T_2(z, 0)$ beneath the block was measured at the depths of 2, 4, 8, 16 and 32 mm. Heat Flux can be obtained from the following expression:

$$H(0,0) = -\lambda_2 E_2 = (1+\alpha)\sqrt{\lambda_1 C_1} \times \frac{\sqrt{\pi}}{2} \times \text{slope of the graph} \qquad (4)$$

where $\lambda_2$ is the thermal conductivity of the sample and $E_2$ is the temperature gradient at different depths.

## 2.2. Sample Preparation

Granite sample slab of approximately 15 cm x 15 cm x 10 cm was cut to size and bored to slightly less than 1mm diameter on one side to allow for close insertion of the KD2 probe. The slab was placed one after the other inside a chamber where the temperature measurement was taken for block method calculation. Component analysis of the granite rock sample using ASTM [17] is shown in Table 1. A Picolog Data Logger (USB TC-08 Thermocouple (USA)) was connected through temperature sensors connected to the block while a 0.05cm-thick TIM was carefully applied to close the clean contact surfaces of the block apparatus and the sample. Arctic Silver® used in this work as the thermal interface material is a high-density compound of silver, aluminum oxide, zinc oxide, and boron nitride in a polysynthetic oil base. It is at least 88 % thermally conductive by weight, while its thermal resistance is rated as less than 0.0045°C-in²/W.

**Table 1. Description of granite sample**

| Samples | Colour | Grain size | Fabric | Mineral contents |
|---|---|---|---|---|
| Granite | Light grey | Block size 15 cm x 15 cm 10 cm | Isotropic | Quartz (30%), microcline (35%), Plagioclase (30%) Others (5%) |

## 2.3. Block with Thermal Interface Material

From the plot of Temperatures against time for granite without TIM, the following calculations were made:

$T_2 = 31.55°C$

$$31.77(1+\alpha) = 31.95 + \alpha T_2 \tag{5}$$

$$31.42(1+\alpha) = 31.30 + \alpha T_2 \tag{6}$$

solving eqns. 5 and 6 simultaneously gives $\alpha = 0.8571$, $\sqrt{\lambda_2 C_2} = 1.1606 \times 10^{-2}$ while $T_{2\,in}$, the sample initial temperature, is $31.55°C$. As the temperature gradient in the upper soil layer is known, we can calculate $\lambda_2$ (thermal conductivity), $C_2$ (volumetric heat capacity) and $\alpha_2$ (thermal diffusivity) separately:

$$\sqrt{a_2} = \frac{\sqrt{\lambda_2}}{\sqrt{C_2}} = \text{slope} \times \frac{\pi}{2} \times \frac{1}{E_2} \times \left(1 + \frac{1}{\alpha}\right)$$

where E is the temperature gradient in the upper soil layer and slope = 0.0527.

$\frac{\delta\theta}{\delta z} = 31.26 - 0.124z$, $E_2 = 0.124$, $\sqrt{a_2} = 0.8161$

Thermal Conductivity ($\lambda_2$) = $\sqrt{\lambda_2 C_2} \times \sqrt{a_2}$ = 3.96 W/m K

Heat flux density (H) = $\lambda_2 \times E_2$ (Thermal gradient) = $4.9 \times 10^{-1}$ W/m²

Volumetric Heat Capacity ($C_2$) = $\sqrt{\lambda_2 C_2} / \sqrt{a_2}$ = $5.95 \times 10^4$ J/m³K

Thermal diffusivity ($\alpha_2$) = $\lambda_2/ C_2$ = 0.67 x $10^{-2}$ $m^2$/s

Results of thermal properties are presented in Table 2, where BO and BW represents Block without and with TIM respectively, while KO and KW represents KD2 without and with TIM respectively.

## 2.4. Block without Thermal Interface Material

From the plot of Temperatures against time for granite without TIM, the following calculations were made:

$T_2$ = 35.01°C

$T_2(0, t)$ = 35.17 - 0.0289 $\sqrt{t}$

$T_2(z, t)$ = 28.14 - 0.089z

$\alpha$ = 0.8125

$\sqrt{\lambda_2 C_2}$ = 1.10025 x $10^{-2}$

$\sqrt{a_2}$ = 0.64201, $E_2$ = 0.089

Thermal Conductivity ($\lambda_2$) = 2.956 W/mK

Heat flux density = $\lambda_2$ x $E_2$ (Thermal gradient) = 2.63 x $10^{-1}$ W/$m^2$

Volumetric heat capacity ($C_2$) = ($\sqrt{\lambda_2 C_2} / \sqrt{a_2}$) = 7.17 x $10^4$ J/$m^3$K

Thermal diffusivity ($\alpha_2$) = $\lambda_2/ C_2$ = 0.4122 x $10^{-4}$ $m^2$/s

## 2.5. Instantaneous Method

Instantaneous measurements were made using a 0.9 mm diameter thermal probe with a KD2 digital meter (Decagon Devices Inc., Pullman, WA, USA) to compare with block measurements for the purpose of validation (Figure2). The thermal probe was considered as an infinitely long heat source in an isotropic medium under a uniform initial temperature [18 – 20]. During measurements, the 60 mm long probe was put into the samples and the in-built microprocessor ensures temperature stability within 90 s before heating the probe for 30 s.

**Table. 2. Evaluation of Block thermal conductivities of the samples**

| Samples | Block Exp. Without TIM W/mK $\lambda_{BO}$ | Block Exp. with TIM W/mK $\lambda_{BW}$ | Difference $\lambda_{BW}-\lambda_{BO}$ | % of difference | KD2 without TIM W/mK $\lambda_{KO}$ | KD2 with TIM W/mK $\lambda_{KW}$ | Difference $\lambda_{KW}-\lambda_{KO}$ | % of difference | Standard range values W/mK (Kappelmayer and Heanel, 1974) |
|---|---|---|---|---|---|---|---|---|---|
| Granite | 2.95 | 3.95 | 1.00 | 25 | 2.93 | 3.96 | 1.03 | 26 | 2.0-7.0 |

Thermal conductivity and Thermal Diffusivity are thus calculated from the temperature rise before instantaneous display of results. KD2 was used to make measurement on the granite sample with and without Thermal Interface Material.

## 3. TRANSIENT METHOD

Fasunwon *et al.* [21], described an experimental technique to measure thermal properties of rocks using a fabricated equipment based on transient theory, and compared results obtained with Kappelmayer and Haenel [22]. The study considered a slab that is initially at zero temperature and insulated at the surface $x = 0$, and has a constant heat flux introduced at the surface $x = a$ at time $t = 0$. Differences observed in the published results were traced to a number of factors: (1) differences in the relative abundances of the mineral compositions and (2) fabrics of the rocks. The accuracy of the thermal diffusivity in the study was strongly related to strict adherence to experimental procedures. The error of the temperature and thickness measurement on the samples was approximated to be from 1-2% and that of thermal conductivity about ±15%.

Figure 2. Drilled hole inside the granite block to allow insertion of the KD2 probe for the measurement.

## 4. DISCUSSIONS

Results indicate that thermal conductivity of granite strongly affects other physical properties as expected [23]. Test of significance level was carried out between the two methods as shown in Table 2, block and KD2 with TIM and without TIM through the analysis of variance (ANOVA) using Fisher's Protected Least Significant Difference. Thermal conductivity determined for granite through block method and KD2 is represented through Figure 3. $f_{cal} = 0.2615$ represents the P-value as greater than 0.05. Results indicate significant difference of thermal conductivities without TIM and with TIM (Table 4). This implies that using TIM to reduce the contact resistance error in the block method is effective. $f_{cal} = 0.24874$ represents the P-value as also greater than 0.05. Analysis indicates significant difference of thermal conductivity without TIM and with TIM (Table 5). Thermal conductivity of granite sample, λ increased from 2.95 to 3.95 W/mK with 25 % difference for the block method and 2.93 to 3.96 W/mK with 26 % difference for the KD2 measurement. Volumetric heat capacity decreased from $7.17 \times 10^4$ J/m$^3$K to $5.95 \times 10^4$ J/m$^3$K, thermal diffusivity increased from $0.412 \times 10^{-4}$ m$^2$/s to $0.67 \times 10^{-4}$ m$^2$/s while heat flux density increased from $2.63 \times 10^{-1}$ W/m$^2$ to $4.9 \times 10^{-1}$ W/m$^2$ (Table 3). Increase in λ values with TIM was significant because of the reduction in thermal contact resistance. The difference between Block measurement and that of KD2 thermal properties analyzer was very insignificant and both fall within the range of the standard values as shown in Table 2.

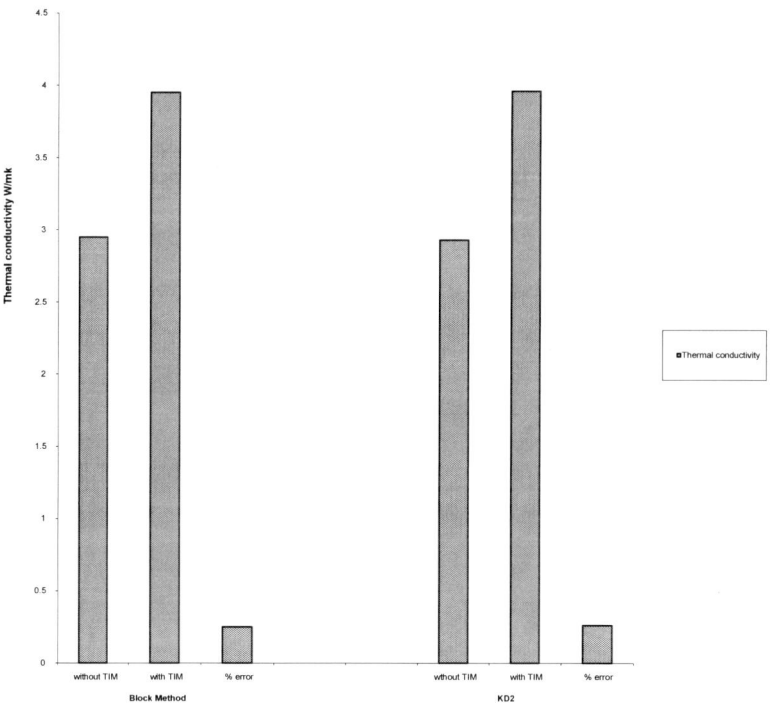

Figure 3. Thermal conductivity of granite without/with TIM.

### Table 3. Calculated thermal properties of granite

| Granite | Without TIM | With TIM |
|---|---|---|
| Thermal conductivities (λ) W/mK | 2.95 | 3.96 |
| Volumetric heat capacity (c) J/m$^3$K | $7.17 \times 10^4$ | $5.95 \times 10^4$ |
| Thermal diffusivity (ρ) m$^2$/s | $0.4122 \times 10^{-4}$ | $0.67 \times 10^{-2}$ |
| Heat flux density (H) W/m$^2$ | $2.63 \times 10^{-1}$ | $4.9 \times 10^{-1}$ |

### Table 4. Statistical Test of Significance for Block Method

| SV | SS | df | MS | $f_{cal}$ |
|---|---|---|---|---|
| With/without TIM | 0.46556125 | 1 | 0.4656125 | 0.2615 |
|  | 10.67948 | 6 | 1.779913 |  |
| Total | 11.1451 | 7 |  |  |

**Table 5. Statistical Test of Significance for KD2 Method**

| SV | SS | df | MS | $f_{cal}$ |
|---|---|---|---|---|
| With/without TIM | 0.44185 | 1 | 0.444185 | 0.24874 |
| | 10.65785 | 6 | 1.7763 | |
| Total | 11.0997 | 7 | | |

## CONCLUSION

Methods of processing thermal properties of granite have been described in this chapter with special emphasis on contact method. Accuracy concerns arising from contact errors have been addressed using thermal interface materials which successfully reduced contact resistance errors, and thus enhanced the effectiveness of block device to determine thermal properties of porous media. Thermal conductivity of granite increased from 2.95 W/mK to 3.95 W/mK, with the standard values for granite ranging from 2.0 and 7.0W/mK.

Volumetric heat capacity decreased from $7.17 \times 10^4$ J/m$^3$K to $5.95 \times 10^4$ J/m$^3$K, thermal diffusivity increased from $0.412 \times 10^{-4}$ m$^2$/s to $0.67 \times 10^{-4}$ m$^2$/s while heat flux density increased from $2.63 \times 10^{-1}$ W/m$^2$ to $4.9 \times 10^{-1}$ W/m$^2$ with application of TIM. Increase in $\lambda$ values with TIM also corresponds with increase in heat flux density and thermal diffusivity but decrease in volumetric heat capacity. Thermal properties without and with TIM showed significant difference at $P > 0.05$, thus confirming the effectiveness of thermal interface materials in addressing contact resistance challenges.

## REFERENCES

[1] O. D. Akinyemi, N. Mendes, Numerical and Experimental Determination of surface temperature and moisture evolution in a field soil, *J. Geophys. Eng.* 4 (2007) 7 – 17.

[2] J. Kim, Y. Lee, M. Koo, Thermal Properties of Granite from Korea, *American Geophysical Union, Fall Meeting 2007, abstract #T11B-0576*

[3] O. D Akinyemi, T. J. Sauer, Y. S. Onifade, Revisiting the block method for evaluating thermal conductivities of clay and granite. *International Communications in Heat and Mass Transfer* 38 (2011) 1014 – 1018

[4] T. J. Sauer, O. D. Akinyemi, P. Thery, J. L., Heitman, T. M., DeSutter, R. Horton, Evaluation of a new, perforated heat flux plate design, *International Communications in Heat and Mass Transfer* 35 (2008) 800 – 804.

[5] W. R. Van Wijk, Two New Methods for the Determination of the Thermal Properties of Soils Near the Surface, *Physica*, 30(2) (1964) 387-388.

[6] W. R. Van Wijk, New Method for Measuring Heat Flux Density at the Surface of Soils or of Other Bodies, *Nature*, 213 (1967) 213-214.

[7] C. J. Stigter, On the Possibility of Determining Thermal Properties from Contact-Surface Temperatures, *Physica*, 39(2) (1968) 229-236.

[8] T. Schneider, The Block Method for Measuring Heat Flux Density at the Surface of Soils and other Solids, *Agricultural Meteorology,* 6 (1969) 423-434.

[9] J. H. Blackwell, A Transient-Flow Method for Determination of Thermal Constants of Insulating Materials in Bulk Theory. *J. Appl. Phy.,* 25(4) (1954) 137-144.

[10] Hadas, Problems Involved in Measuring the Soil Thermal Conductivity and Diffusivity in a Moist Soil, *Agricultural Meteorology,* 13(1) (1974) 105-113.

[11] Hadas, Heat Transfer in Dry Aggregated Soil: Heat Conduction, *Soil Sci. Soc. Am. J.,* 4 (1977) 1055-1059.

[12] T. J. Sauer, T. E. Ochsner, R. Horton, Soil Heat Flux Plates: Heat Flow Distortion an Thermal Contact Resistance, *Agronomy Journal,* 99(1) (2005), 304-310.

[13] A. Van Haneghem, J. Schenk, H. P. A. Boshoven, An Improved Nonsteady-State Probe Method for Measurements in Granular Materials. Part 2: *Experimental Results, High Temperatures – High Pressures,* 15 (4) (1983) 367-374.

[14] N. K. Nagpal, L. Boersma, Air Entrapment as a Possible Source of Error in the Use of a Cylindrical Heat Probe, *Soil Sci. Soc. Am. Proc.,* 37 (1973) 828-832.

[15] G. S. Campbell, E. M. Huffaker B. T. Wacker, K. C. Wacker, Use of the Line Heat Source Method to Measure Thermal Conductivity of Insulation and other Porous Materials. *Thermal Conductivity Decagon Devices, Inc., 950 NE Nelson Ct., Pull man, WA 99163, USA, 2003.*

[16] H. S. Carslaw, J. C. Jaeger, *Conduction of Heat in Solids*, $2^{nd}$ ed., Clarendon Press, Oxford,UK, (1959) 426-428.

[17] American Society for Testing and Materials (ASTM) D5334-00, Standard Test Method for Determination of Thermal Conductivity of Soil and Soft Rock by Thermal Needle Procedure, in: *Annual Book of ASTM Standards* 4, 2000, 8.

[18] M. Fuchs, A. Hadas, Analysis and Performance of an Improved Soil Heat Flux Transducer, *Soil Sci. Soc. Am. Proc.,* 37 (1973) 173-175.

[19] T. J. Sauer, D. W. Meek, T. E. Ochsner, A. R. Harris, R. Horton, Errors in Heat Flux Measurement by Flux Plates of Contrasting Design and Thermal Conductivity, *Vadoze Zone J.,* 2(4) (2003) 580-588.

[20] P. J. Bruijn, I. A. Van Haneghem, J. Schenk, An Improved Nonsteady-State Probe Method for Measurements in Granular Materials, Part 1: *Theory, High Temperatures – High Pressures,* 15(4) (1983) 359-366.

[21] O. O. Fasunwon, J. A. Olowofela, O. O. Ocan, O. D. Akinyemi, Determination of Thermal Conductivity of Rocks samples using fabricated equipment. *Thermal Science:* vol. 12 (2008), No. 2, 119-128.

[22] O. Kappelmayer, R. Haenel, Geothermics with Special References to Application, *Gebruder Borntraegen,* Belin, Stuttgart, (1974) 1-23 and 203-222, 197.

[23] Popov, Y. A., Tertychnyi, V., Romushkevich, R., Korobkov, D., and Pohl, J. Interrelations between thermal conductivity and other physical properties of rocks: *experimental data, Pure app. geophys.*(2003), 160, 1137-1161.

*Chapter 8*

# THE SHOSHONITIC GRANITOIDS OF ALTAI-SAJAN FOLDED AREA: PETROLOGY AND ORE MINERALIZATION

## *A. I. Gusev*[1*] *and N. I. Gusev*[2]

[1]Biisky pedagogical state university, Biisk, Russia
[2]A.P. Karpinsky Russian Geological Research Institute,
Saint-Petersberg, Russia

## ABSTRACT

In the paper presented data about shoshonitic granitoids Altai-Sajan folded area. Petrology of granitoids and link with them ore mineralization considered. Deposits of W, Mo, Ta, Nb, Au, Be, REE known in paragenesis and space link with shoshonitic granites.

**Keywords:** Shosonitic granitoids, ore mineralization, petrology, fluid regime

For the first time the shoshonitic type granitoids (SH) derived China's investigaters in during study intrusive bodies N-W part of China (Jiang, Jiang et.all, 2002). The shoshonitic group of granitoids incorporate assemblages monzogabbro - monzodiorite - monzonite – quartz syenite, or monzonite granite – granite, or biotite (monzonite) granite - diopside granite – diopside syenite. Its type granitoids described by us in Altai-Sajan area (Altai Mountain) and classified to post collisional setting, initiation by function Siberian superplum (Gusev, Gusev, Tabakaeva, 2008; Gusev, 2010).

The shoshonitic granitoids founded on Altai-Sajan region in much sites: Savvushinskiy (Rudnyi Altay), Aiskiy, Terandjikskiy, Tarchatinskiy areals (Gornyi Altai), Zhernovskoi, Borsukskiy, Gornovskoy (Salair), Saksyrskiy (Sajan), Askizskiy (Batenevskiy krjadg), Borok-Bibeevskiy (Tom-Kolyvanskaja zone), Beloiussko-Tuimskiy (Kuzneckiy Alatau). It type granitoids and of its areals occurred in the edges of stocks (square 2 - 96 km$^2$) that has

---

[*] Russia, Email: anzerg @ mail. ru.

composite composition from monzogabbro to leicogranites. The stable paragenesis of dikes different composition from dolerites to granites with lamprophyres and massifs with appinites are watched in all areals. The lamprophyres are varies on different types of rocks (spessartites, vogesites, minetts, kersantites), but minetts occur in all areals from mafic to felsic types, that its relate to alkaline-basaltic of mantle magmas.

The shoshonitic granitoids are characterized by contents $SiO_2$ from 52.77 to 71.85% and high sum alkali $K_2O+Na_2O$ (more > 8%, average 9.14%), ratio $K_2O/Na_2O$ (more >1, average 1.50) and ratio $Fe_2O_3/FeO$ (0.85–1.51, average 1.01) and low content $TiO_2$ (0.15–1.12%, average 0.57%). It type granitoids are characterized by high conctntrations Ba and Sr. J. Tarney and C. Jones (1994) drew specific attention to these elements, together with low Rb and consequent high K/Rb, low Th, U and Nb, and very low Y and heavy REE relative to other trace elements, the combination of which defines the high Ba-Sr granitoid group in Scotland. Before L.V. Tauson separated latite geochemical type granitoid (Tauson, 1977), that it is correspond shosonite and high Ba-Sr types.

The content of $Al_2O_3$ in rocks vary from 13.01 to 19.20% and very variable. The granitoids enrich by LILE, LREE and volatile components, such as F, B and other. The biotite of shoshonitic granitoids classify to ferruginous phlogopite with minor fraction estonite and high ratio $Mg/(Mg + Fe_t)$ and $Fe_3+/Fe_2+$. Amphibole classify to edenite hornblende and magnesian hastingsite with some fraction edenite and high ratio $Mg/(Mg + Fe_t)$ and $Fe_3+/Fe_2+$.

The representable analysis of rocks some intrusive massifs with shoshonitic granitoids tabulate in table 1.

The analysis complete: for main components – by chemical method, for elements- by method ICP-ms IMGRE (c. Moscow). Salair: Gornovskoy massif :1- granite; Zhernovskoi massif: 2- syenite; Gornyi Altai: Terandjikskiy massif: 3- granite; Gornyi Altai: Aiskiy massif: 4- syenite, 5-granosyenite, 6- granite, 7- leicogranite, 8- leicogranite with fluorite.

The under study rocks of region fall on diagram (on composite of biotites) in field of shoshonitic granitoids (figure 1).

A potential ore mineralization of intrusive complexes and separate bodies can be determine by path of calculation rare metal index – F(Li+Rb)/(Sr+Ba) to L.V. Tauson (1977) with account of distinction fluid regime and concentration of volatile components in it (F, $H_2O$, B). The values of rare metal index and other necessary dates on example Aiskiy massif actuation in table 2. Analysis of table 2 show that appreciable increase concentration F and rare metal index occur from monzogabbro to leicogranite. The values of rare metal index (6178,3) and petro-geochemical parameters are very closely to peraluminous rare metal leicogranites (rare metal index 6800). The analogous parameters for leicogranite with fluorite, that it is paragenetic connect greisens and pegmatites deposits of Sn, Ta, Nb under investigation region.

f – total mafic index of biotites (f= Fe+Mn/Fe+Mn+Mg); L – aluminous of biotites (L= Al/Si+Al+Fe+Mg); OH/F – ratio hydroxyle group to fluorine in composite biotites. Standard type granitoids: M- mantle MOR, backarc basins (in composition of ophiolite complexes); I- mantle-crustal of island arc, transform, active continental margins, collision settings; S – crustal and mantle-crustal of collision settings and complexes metamorphic cores; SH- shoshonitic type granitoids of post collision settings, initiating by plum tectonic; A- mantle-crustal and mantle of anorogenic settings (intracontinental rifts, hot spots, mantle plumes).

**Table 1. The representable analysis of shoshonitic granitoids some intrusive massifs (main components in %, elements – ppm)**

| Components | 1 | 2 | 3 | 4 | 5 | 6 | 7 | 8 |
|---|---|---|---|---|---|---|---|---|
| $SiO_2$ | 70.31 | 66,31 | 71,97 | 61,87 | 66,11 | 72,87 | 75,05 | 76,88 |
| $TiO_2$ | 0.42 | 0,49 | 0,17 | 1,20 | 0,47 | 0,16 | 0,13 | 0,11 |
| $Al_2O_3$ | 14.08 | 16,44 | 14,16 | 17,28 | 16,64 | 13,96 | 13,67 | 12,92 |
| $Fe_2O_3$ | 2.09 | 1,39 | 0,72 | 2,12 | 1,44 | 0,75 | 0,56 | 0,37 |
| FeO | 1.10 | 1,35 | 0,81 | 2,01 | 1,37 | 0,83 | 0,65 | 0,36 |
| MnO | 0.06 | 0,08 | 0,04 | 0,12 | 0,09 | 0,04 | 0,03 | 0,03 |
| MgO | 1.10 | 1,11 | 0,33 | 0,67 | 1,01 | 0,37 | 0,22 | 0,11 |
| CaO | 2.65 | 2,13 | 0,59 | 2,12 | 2,10 | 0,49 | 0,59 | 0,32 |
| $Na_2O$ | 3.82 | 4,91 | 4,65 | 3,04 | 4,89 | 4,61 | 3,89 | 4,09 |
| $K_2O$ | 3.58 | 5,15 | 4,62 | 8,95 | 5,12 | 4,72 | 4,65 | 3,93 |
| П.п.п | 0,05 | 0,21 | 0,31 | 0,40 | 0,23 | 0,32 | 0,42 | 0,41 |
| $P_2O_5$ | 0,58 | 0,18 | 0,05 | 0,16 | 0,16 | 0,05 | 0,03 | 0,03 |
| $\sum$ | 99.23 | 99,73 | 99,27 | 99,96 | 99,63 | 99,17 | 99,89 | 99,56 |
| Li | 43.1 | 27.5 | 30.4 | 18,8 | 27,6 | 55 | 4,5 | 10,8 |
| Rb | 107,2 | 94,2 | 125,5 | 109 | 78,9 | 145 | 164 | 172 |
| Cs | 2,4 | 2,4 | 3.1 | 2,2 | 2,8 | 3,6 | 7,5 | 1,9 |
| Be | 5,1 | 1.5 | 5.5 | 3,8 | 0,7 | 5,3 | 6,7 | 0,7 |
| Sr | 1063 | 2520 | 2200 | 8750 | 630 | 2280 | 20 | 8 |
| Ba | 1100 | 1990 | 2500 | 1956 | 750 | 2310 | 40 | 20 |
| La | 47,0 | 66 | 47 | 46 | 73 | 74 | 55 | 32 |
| Ce | 69,3 | 74 | 86 | 58 | 86 | 97 | 63 | 36 |
| Nd | 28,2 | 22 | 25 | 24 | 24 | 29 | 16 | 8,6 |
| Sm | 5,67 | 4.6 | 5.5 | 5,4 | 4,2 | 5,5 | 2,2 | 0,9 |
| Eu | 1,44 | 1.37 | 1.64 | 1,42 | 1,23 | 1,21 | 0,68 | 0,13 |
| Gd | 5,0 | 3.6 | 4.5 | 6,1 | 3,3 | 4,1 | 2,1 | 0,9 |
| Tb | 0,73 | 0.9 | 1.11 | 0,94 | 0,52 | 0,61 | 0,26 | 0,11 |
| Dy | 1.21 | 2.3 | 4.1 | 3,9 | 2,3 | 1,2 | 1,6 | 0,7 |
| Tm | 0,3 | 0.3 | 0.5 | 0,4 | 0,3 | 0,2 | 0,2 | 0,2 |
| Yb | 1.18 | 2.4 | 3.1 | 2,8 | 1,22 | 1,6 | 1,1 | 1,4 |
| Lu | 0,16 | 0.3 | 0.5 | 0,4 | 0,3 | 0,25 | 0,21 | 0,2 |
| Y | 9,6 | 11.8 | 13.7 | 14,7 | 7,8 | 13,6 | 13,4 | 10,4 |
| Sc | 4,3 | 5.7 | 6.5 | 5,7 | 5,6 | 4,2 | 3,3 | 1,3 |
| Th | 1.56 | 4.5 | 15.8 | 5,4 | 24 | 27 | 41 | 48 |
| Hf | 4,6 | 4.8 | 4.9 | 18 | 5,2 | 6,9 | 4,6 | 4,6 |
| Ta | 0,6 | 1.5 | 0.5 | 0,9 | 1,66 | 3,2 | 2,2 | 4,8 |
| Nb | 3,2 | 5.2 | 6.3 | 22,7 | 35,3 | 87,6 | 85,2 | 77 |
| Zr | 221 | 318 | 334 | 276 | 243 | 238 | 204 | 215 |

The shoshonitic granitoids of massif region: 1- Aiskiy, 2- Gornovskoy, 3- Terandjikskiy, 4- Tarchatinskiy, 5- Tarchatinskiy, 6- Beloiussko-Tuimskiy.The intrusive massifs of shoshonite granitoids in Altai-Sajan region, with it connect W-Mo skarns (Plitninskoe, Aturkolskoe deposit), W-Mo greisens (Orlinoe, Osokinskoe, Osinovskoe deposits) and lode W-Mo deposits (High-Belokurikchinskoe, Dmitrievskoe, Batunkovskoe), lode Be (Kazandinskoe), pegmatitic and greisens beryllium (Kuranovskoe), Ta-Nb (River Slepoy), Li deposits, so lode gold-sulfide-quartz (Atbashi) manifestations. The skarns deposits complex W-Sn with Au known in the contact of massif Karagu (West Karagu, East Karagu).

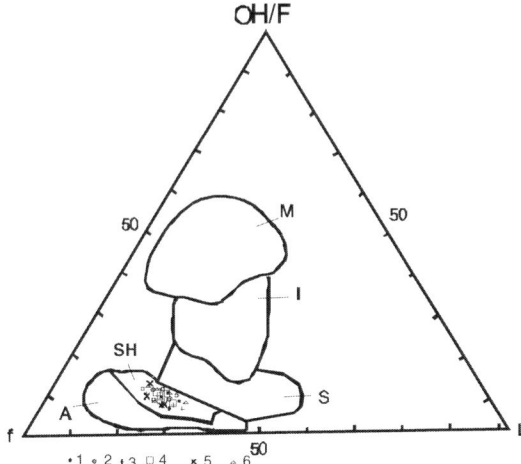

Figure 1. Diagram f- L- OH/F in biotites of granitoids.

**Table 2. Concentration rare elements and values of rare metal index in rocks of Aiskiy massif**

| Rocks | F,% | Li, г/т | Rb, г/т | Sr, г/т | Ba, г/т | F(Li+Rb)/(Sr+Ba) |
|---|---|---|---|---|---|---|
| Monzogabbro | 0,02 | 21,2 | 95 | 1950 | 2070 | 5,78 |
| Monzonite | 0,03 | 20,5 | 104 | 2720 | 1970 | 7,96 |
| Melanosyenite | 0,04 | 30,1 | 125 | 2200 | 2500 | 13,2 |
| Syenite | 0,08 | 18,8 | 109 | 8750 | 1956 | 9,54 |
| Granosyenite | 0,10 | 27,6 | 78,9 | 630 | 750 | 77,2 |
| Granite | 0,12 | 55 | 145 | 280 | 310 | 406,8 |
| Leicogranite | 0,22 | 4,5 | 164 | 20 | 40 | 6178,3 |
| Leicogranite with fluorite | 0,85 | 10,8 | 172 | 7 | 20 | 57548,1 |

The pegmatitic deposit Ortitovoe confine in Savvushinskiy massiv of shoshonitic granitoids in Rudnyi Altai. The allanite is main ore mineral on it pegmatite. Allanite occur in disseminate form and it form large crystals to 30 sm in length. It associated with schörl, albite, muscovite, rauhtopaz.

The complex deposit Au-U-W confine in the contact of intrusive Tarchatinskiy areal of shoshonitic granitoids (Gornyi Altay). The lodes of deposit Elangash contain nasturane, gold, scheelite, wolframite.

In the final report follow to say, that the shoshonitic granites of Altai-Sajan region derived in compound post collision setting, initiation by function Siberian superplum and it characterized by saturate volatile components (F, $H_2O$, B). There are define of it potential ore mineralization. The different deposits and manifestations of W, Mo, Ta, Nb, Au, Be, REE known in paragenesis and space link with shoshonitic granites in Altai-Sajan region (Kuzneckiy Alatau, Salair, Sajan).

## REFERENCES

Gusev A.I., Gusev N.I., Tabakaeva E.M. The petrology and ore mineralization of Belokurikhinskii complex of Gornyi Altai. *Biysk*, 2008. – 193 p.

Gusev A.I. Minerageny and mineral resources of Republic Altai. *Biysk*, 2010. – 395 p.

Jiang Y-H, Jiang S-Y, Ling H-F, Zhou X-K, Rui X-J, Yang W-Z. Petrology and geochemistry of shoshonitic plutons from the western Kunlun orogenic belt, Xinjing, northwestern Chine: implications for granitoids genesis// *Lithos*, 2002.- V.63.- P. 165-183.

Tarney J., Jones C.E. Trace element geochemistry of orogenic igneous rocks and crustal growth models //*Journal Geol. Soc.*, London, 1994.- V. 151.- P. 855-868.

Tauson L.V. The geochemical types and carrying ore mineralization of granitoids. - M.: - *Science.* – 1977. - 280 p.

# INDEX

## A

accounting, 43
acid, ix, 17, 57, 66, 80, 91
acoustics, 129
actuation, 144
AFM, 85
Africa, ix, 58, 59, 91, 94, 95, 96
age, vii, x, 1, 2, 4, 7, 10, 16, 17, 22, 26, 61, 62, 65, 75, 79, 80, 90, 91, 95
agriculture, 132
Algeria, 82
aluminum oxide, 135
amplitude, xi, 117, 118, 119, 120, 121, 122, 123, 124, 125, 126
ASI, 28
assessment, 108
assimilation, 59, 61, 65, 67, 68, 69, 70, 74
atoms, 33
Austria, 61, 66

## B

banks, 18
base, 135
Bastar Cratonic Terrain (BCT), viii, 2, 4, 10, 11, 22
beryllium, 17, 145
bicarbonate, 111
boreholes, 132
Brazil, 95
breakdown, 92
Brno, 33, 54

## C

$Ca^{2+}$, 43
calcium, 49
Cameroon, 82, 83
catalyst, 2
Central Europe, viii, x, 27, 28, 53, 99, 100, 101, 113
cerium, 18, 21, 25, 26
Chad, 80
challenges, 140
chemical, vii, 1, 2, 7, 11, 14, 22, 23, 44, 50, 67, 77, 103, 110, 144
Chhotanagpur Granite Gneiss Complex (CGGC), vii, 1, 3, 4, 13, 17, 21
China, 25, 43, 143
classification, vii, 1, 14, 21, 23, 60, 81, 84, 94, 95, 106, 107, 108
cleavage, 5, 8
closure, 83, 95
clusters, 92
$CO_2$, 111
collage, 28
color, 119
commercial, x, 80
compilation, vii
composition, viii, x, 2, 3, 5, 6, 7, 8, 9, 11, 14, 15, 19, 20, 28, 33, 53, 55, 56, 58, 62, 65, 66, 67, 69, 70, 77, 79, 83, 93, 100, 101, 102, 110, 111, 115, 144
compounds, 133
conduction, 133
conductivity, xi, 131, 132, 133, 134, 135, 138, 139, 140, 141
conference, 94
conformity, 15
Congo, 80, 82
consensus, 58
constituents, 6, 10, 13, 132
consumption, 59
contamination, ix, 58, 65, 67, 68, 69, 70, 72, 73, 74, 77, 85
Continental, 83, 95
controversial, 108
convergence, ix, 58, 59

cooling, 14
correlation, 24, 25, 47, 49, 62, 65, 94, 112
correlations, 65
covering, 133
cracks, xi, 5, 117, 118
Croatia, 74
crude oil, x, 80
crust, viii, ix, 2, 17, 22, 55, 58, 59, 62, 65, 66, 67, 68, 69, 70, 73, 76, 78, 90, 91, 115
crustal growth, 147
crystallisation, vii, 1, 13, 15, 21, 59, 65, 69
crystallization, 28, 59, 61, 65, 66, 69, 91, 107, 111
crystals, 14, 20, 33, 39, 49, 93, 146
cycles, 15, 81
Czech Republic, 27, 30, 33, 99, 100, 102, 113, 115

## D

damping, 120, 122, 124, 126, 127, 128, 129
decoupling, 68, 91
defects, xi, 117, 118
deficit, 39
deformation, vii, 1, 3, 15, 23, 81, 85, 96, 118, 128
dehydration, 110, 111
Delta, 80
deposits, x, 2, 18, 21, 23, 25, 29, 56, 58, 80, 93, 94, 115, 144, 145, 146
depth, 14, 17, 18, 32, 101, 133
detachment, 76
detection, 43, 47, 48, 49, 53
deviation, 120
dichotomy, 100
diffusivity, xi, 131, 132, 133, 135, 136, 138, 139, 140
diorites, x, 60, 61, 66, 68, 79
discrimination, 96, 107, 109, 113
dislocation, xi, 117, 121
dispersion, 127
displacement, 121, 123, 124
disposition, 90
dissipative nonlinearities, xi, 117, 119
distribution, ix, 16, 29, 47, 49, 53, 58, 69, 88, 90, 118
dykes, ix, x, 17, 58, 60, 61, 62, 65, 74, 79, 80, 84, 86, 87, 91, 93, 100

## E

Eastern Europe, 55
economics, 26
electron, 32, 43
energy, 132
engineering, 132

England, 76
environment(s), 15, 84, 93
equipment, 138, 141
erbium, 25
Europe, ix, 58, 59
europium, 25
evidence, ix, 58, 62, 65, 67, 69, 74, 75, 76, 77, 78, 86, 91, 92, 95
evolution, ix, 15, 23, 24, 25, 40, 55, 58, 59, 62, 65, 67, 73, 74, 80, 83, 91, 94, 95, 110, 111, 115, 140
excitation, 120, 123, 125, 126, 127
exposure, 28, 29
extinction, 5, 8, 85
extraction, 112

## F

facies, 3, 83, 100, 104
fertility, 96
fine-grained grey granitoid (FGGG), vii, 1, 21
flank, 100
fluid, 29, 74, 111, 143, 144
fluorescence, 33, 102
fluorine, 36, 49, 100, 104, 106, 111, 112, 114, 144
formation, xi, 13, 15, 23, 92, 99, 111
formula, 33, 40, 43, 120
fractures, 4, 91
fragments, 20
France, 77
friction, xi, 117, 118
fusion, 33, 104

## G

geography, 95
geology, 24, 55, 75, 83, 95, 96, 97
geometry, 74
Germany, 30, 100, 115
global scale, 59
grades, 3
grading, 13, 21
grain boundaries, xi, 117
grain size, 5, 6, 8, 14, 15
granite sample, xi, 32, 121, 125, 131, 132, 135, 138, 139
granules, 119
graph, 134
grey granitoid (GG), vii, 1, 6, 21
growth, 29, 86

## H

hafnium, 48, 49
heat capacity, xi, 131, 132, 133, 135, 136, 139, 140
heat transfer, 132
heavy-rare-earth-element (HREE), vii, 2
heterogeneity, 70, 74, 76
history, 15, 24, 53, 77
host, x, 17, 22, 54, 79, 85, 86, 99
hot spots, 97, 144
hybrid, 66, 68, 78, 91
hydrothermal activity, 10, 14
hydrothermal process, viii, 27, 93
hypothesis, 133
hysteresis, xi, 117, 118, 123, 124, 128

## I

ideal, 43
identification, 94, 110
IMA, 40
India, v, vii, 1, 2, 5, 6, 7, 8, 9, 11, 12, 16, 17, 19, 20, 21, 22, 23, 24, 25, 26
initiation, 91, 143, 146
insertion, 134, 138
insulation, 133
interface, xi, 131, 135, 140
intrusions, 23, 29, 59, 60, 64, 65, 66, 67, 69, 74, 76, 101
Iowa, 131
iron, 5, 7, 49, 85, 92
isotope, ix, 58, 61, 62, 64, 65, 66, 67, 68, 69, 70, 73, 74, 75, 76, 110
isotopic ratios, ix, 58, 61
Italy, v, 57, 74, 75, 76, 77, 78

## J

joints, 4

## K

Korea, 132, 140

## L

lanthanide, 112
lead, 75
light, 13, 16, 17, 23
line graph, 134

linear dependence, 120, 121
lithium, x, 33, 35, 39, 99, 104
local conditions, 74
Luo, 55
lying, 4, 101

## M

magnitude, 128
majority, viii, 2, 12, 27, 29, 30, 53, 87
man, 141
mantle, ix, 58, 59, 61, 62, 64, 65, 66, 67, 68, 69, 70, 71, 72, 73, 74, 75, 76, 77, 85, 90, 91, 97, 111, 144
mapping, 94, 97
mass, 33, 67, 80, 102, 132, 133
materials, xi, 14, 69, 110, 117, 118, 123, 128, 132, 140
matrix, 33
matter, 29, 107, 109, 113
measurement(s), 86, 119, 125, 132, 133, 134, 136, 138, 139
media, 118, 132
Mediterranean, 75, 76
melt, vii, viii, xi, 1, 2, 13, 14, 15, 16, 17, 21, 22, 27, 28, 47, 76, 99, 107, 111
melting, viii, xi, 27, 29, 53, 61, 62, 65, 73, 74, 77, 99, 110, 111
melts, vii, viii, xi, 2, 3, 14, 15, 21, 27, 53, 62, 65, 66, 67, 68, 69, 70, 72, 77, 99, 100, 106, 110, 111, 112
metallurgy, 23
metals, x, 2, 23, 79, 100
microscope, 4, 6, 8, 9
microstructure, xi, 117
migration, 96
mineral resources, 2, 18, 95, 96, 147
mineralization, x, xii, 17, 22, 31, 79, 80, 86, 87, 90, 93, 94, 95, 96, 97, 99, 100, 101, 143, 144, 146, 147
Ministry of Education, 54
Miocene, 58, 60, 69, 72, 73, 74, 76
mixing, 59, 66, 67, 68, 69, 70, 73
modal compositions, vii, 1, 13, 14
modelling, 67, 68, 69, 77
models, 67, 69, 70, 83, 111, 118, 147
modifications, 58, 67, 74
modulus, 121, 122
moisture, 132, 140
Moldanubian Zone, viii, 27, 29, 53
Moscow, 129, 144

## N

Nd, ix, 20, 33, 40, 41, 43, 45, 47, 58, 61, 62, 65, 67, 68, 74, 75, 76, 77, 78, 90, 110, 112, 145
Nigeria, v, vii, ix, 23, 79, 80, 81, 82, 83, 84, 85, 86, 87, 89, 90, 93, 94, 95, 96, 97, 131
niobium, 22, 26
North Africa, 58, 73

## O

oil, 135
ore mineralization, xii, 31, 90, 101, 143, 144, 146, 147
ores, 85, 92
orogenic magmatic rocks, ix, 57
overlap, 107
oxygen, 15, 28, 33, 64, 73, 75

## P

parallel, 4, 64, 84, 85
partition, 73, 107
petroleum, 2, 131
phenomenological state equation, xi, 117, 119, 121, 127
phosphorus, x, 36, 54, 99, 104, 111, 114
physical properties, 132, 138, 141
pink granitoid (PG), vii, 1, 21
plastic deformation, 84
porosity, 132
porous materials, 132
porous media, xi, 131, 132, 140
porphyritic granitoid (PRG), vii, 1, 4, 21
positive correlation, 48, 49, 64, 65, 67, 112
potassium, 35, 36, 64, 96, 104, 105, 106
preparation, 23
principles, 108, 118
probe, xi, 43, 131, 134, 136, 138
project, 54, 94, 113
propagation, xi, 117, 125
P-value, 138

## Q

quantification, 112
quartz, viii, x, 2, 3, 4, 5, 6, 8, 9, 10, 11, 13, 17, 33, 34, 35, 36, 62, 64, 65, 70, 79, 80, 83, 84, 85, 86, 87, 92, 93, 94, 99, 104, 105, 106, 110, 111, 119, 135, 143, 146

## R

radiation, 5
rare-earth-element (REE)-bearing granitoids, vii, 1, 24
recognition, 107
researchers, xi, 117
resistance, xi, 131, 132, 138, 140
resolution, 74
resonator, xi, 117, 119, 120, 121, 123, 124, 126, 127, 128
resources, 17, 20, 23
response, vii, 1, 15, 124
rings, 87
river basins, 3, 18
room temperature, 119
root, 110, 132
Royal Society, 76
rules, 107
Russia, 117, 143
rutile, 4, 18, 36, 105, 106

## S

saturation, viii, 27, 28, 34, 36, 104
sediments, 65, 73, 80
sensors, 134
shear, 25
shoshonitic granites, xii, 143, 146
showing, 29, 63, 69, 70, 73, 88
significance level, 138
silica, 8, 10, 13, 14, 15, 92
silver, 135
$SiO_2$, 5, 7, 8, 9, 10, 11, 12, 14, 15, 19, 20, 28, 41, 44, 50, 60, 61, 64, 65, 85, 103, 110, 144, 145
solid state, 85
solubility, xi, 99, 111
solution, viii, 28, 43, 47, 53, 133, 134
South Africa, 87
South America, 91
South Bohemian batholith (SBB), viii, 27, 53
specific heat, 133
stability, 107, 136
standard deviation, 18
state, xi, 117, 118, 119, 121, 124, 125, 127, 143
steel, 125
stock, 30, 32, 39, 43, 49, 53, 62, 65, 100, 102, 104, 105, 112
stress, ix, 58, 73, 74, 118, 122
strontium, 75
structure, xi, 4, 24, 31, 33, 64, 84, 91, 99, 100, 107, 117, 118

style(s), 73, 93
substitution(s), ix, 28, 43, 47, 48, 54
succession, 91
Sun, 72, 77
superconductivity, 2
surface layer, 94, 133
suture, 96

## T

talc, 3
tantalum, 22, 23, 26
TAP, 33
technical assistance, 54
temperature, 3, 15, 112, 132, 133, 134, 135, 136, 138, 140
tetrad, 112
texture, 4, 6, 7, 8, 9, 10, 13, 14, 15, 87
thermal properties, vii, xi, 131, 132, 133, 136, 138, 139, 140
thermal resistance, 135
thinning, 75
thorium, 26
tin, 17, 22, 23, 24, 25, 86, 91, 95, 110
Togo, 82
tonalites, x, 60, 62, 66, 76, 79, 93
tourmaline, x, 3, 79, 86, 90
trace elements, 33, 64, 77, 107, 144
trajectory, 68
treatment, 112
tungsten, 110
turtle, 84
twinning, 5

## U

UK, 141
UNESCO, 94

uniform, 118, 132, 133, 136
uranium, 25, 49
USA, 20, 21, 78, 131, 134, 136, 141
USSR, 129

## V

validation, 69, 74, 136
variations, ix, 7, 13, 14, 57, 58, 59, 61, 67, 69, 70, 73, 84
varieties, 34, 89, 100
vein, 3, 5, 84
velocity, 120, 125
Vietnam, 26
viscosity, 123

## W

Wales, 94
water, xi, 33, 39, 43, 117, 132
West Africa, 80, 82, 83, 94
workers, 83, 91

## Y

yield, 122
yttrium, 2, 18, 21, 23, 25, 26

## Z

zinc, 95, 135
zinc oxide, 135